Fields and Pine Trees

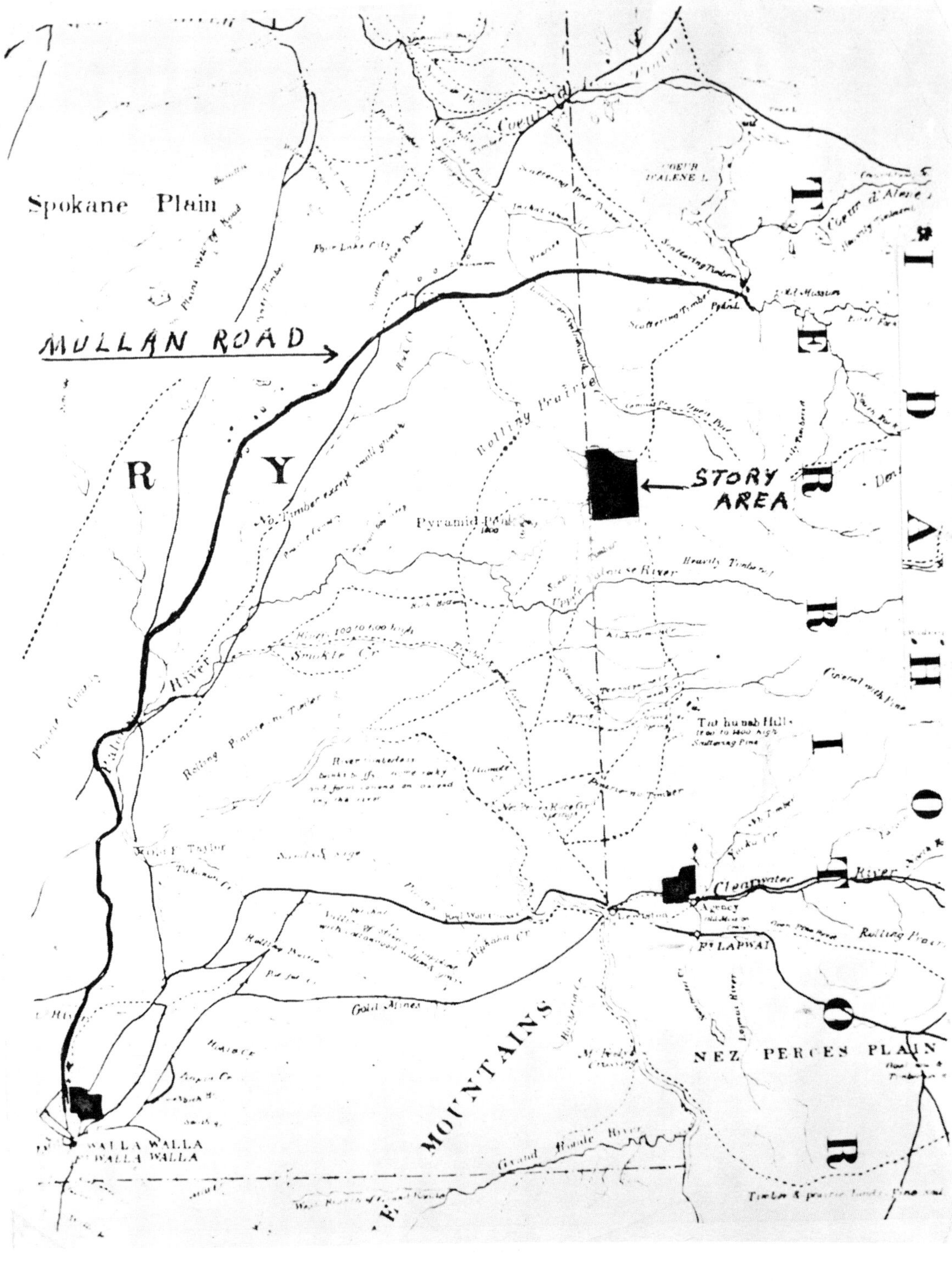

Map No. 1. Territorial area along Washington-Idaho line, showing story area.

FIELDS & PINE TREES

By
Opal Lambert Ross

YE GALLEON PRESS
Fairfield, Washington
1976

Dedicated to my children, Anita and David, who grew up in Evergreen.

Library of Congress Cataloging in Publication Data

Ross, Opal Lambert, 1906-

Fields & pine trees.

1. Latah Co., Idaho--History. 2. Frontier and pioneer life--Idaho--Latah Co. 3. Latah Co., Idaho--Genealogy. I. Title.

F752.L3R67 979.6'86 77-4138

ISBN 087770-177-6

Local History Paper No. 2

Latah County Museum Society

Moscow, Idaho

Editor's Note

The tide of pioneer settlers that swept into the Palouse Country a hundred years ago came from the west. At first a backwash from the Willamette Valley in Oregon, filled by Oregon Trail emigrants in the decades since 1840, the tide soon picked up newcomers where it met them on the Trail at Walla Walla. Its main current followed up the Mullan military road laid out in 1858 along the old Walla Walla-Spokane Indian trail up the Palouse River drainage, the main gateway until the railroad's provided new avenues in the late 1880s.

The first lappings of this tide reached Idaho in 1869-70, touching at Pine Creek in the north end of what is now Latah County, at Kennedy Ford and Moscow in the middle, and at Genesee at the south end. The account given by Mrs. Ross covers the Pine Creek community. There, on January 1, 1872, Lyman Davenport made the first homestead entry registered in Latah County.

This community soon divided itself into three school districts; Evergreen, Pigeon Hollow, and Fairview. In adjoining Washington, but touching the Idaho line where Pine Creek flowed across it, sprang up the town of Farmington.

Destined to brief glory as a rail terminus with dreams of trunk line connections to the east, Farmington found its solid substance as the supply center for rich agricultural lands that drew a family for every 160 acres in the settlement period. Later the town ebbed as the settlement tide ebbed.

The three adjacent school districts closely reflected these epic developments. Mrs. Ross ably tells their story in a microcosmic record of the Pioneer period. Her tracing of each homestead down to its present ownership is a further service to history, as well as a benchmark for possible future study.

Kenneth Platt

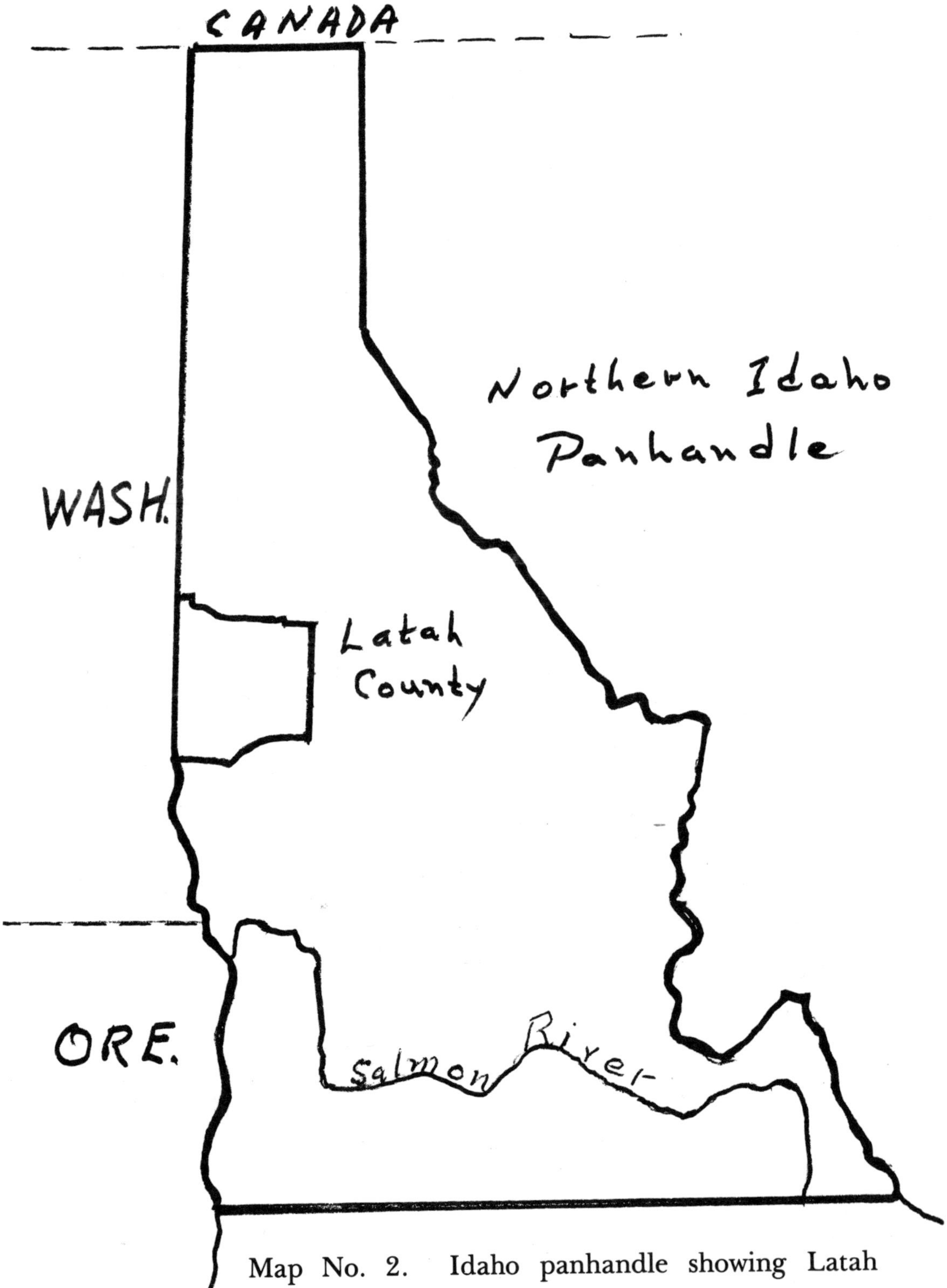

Map No. 2. Idaho panhandle showing Latah County.

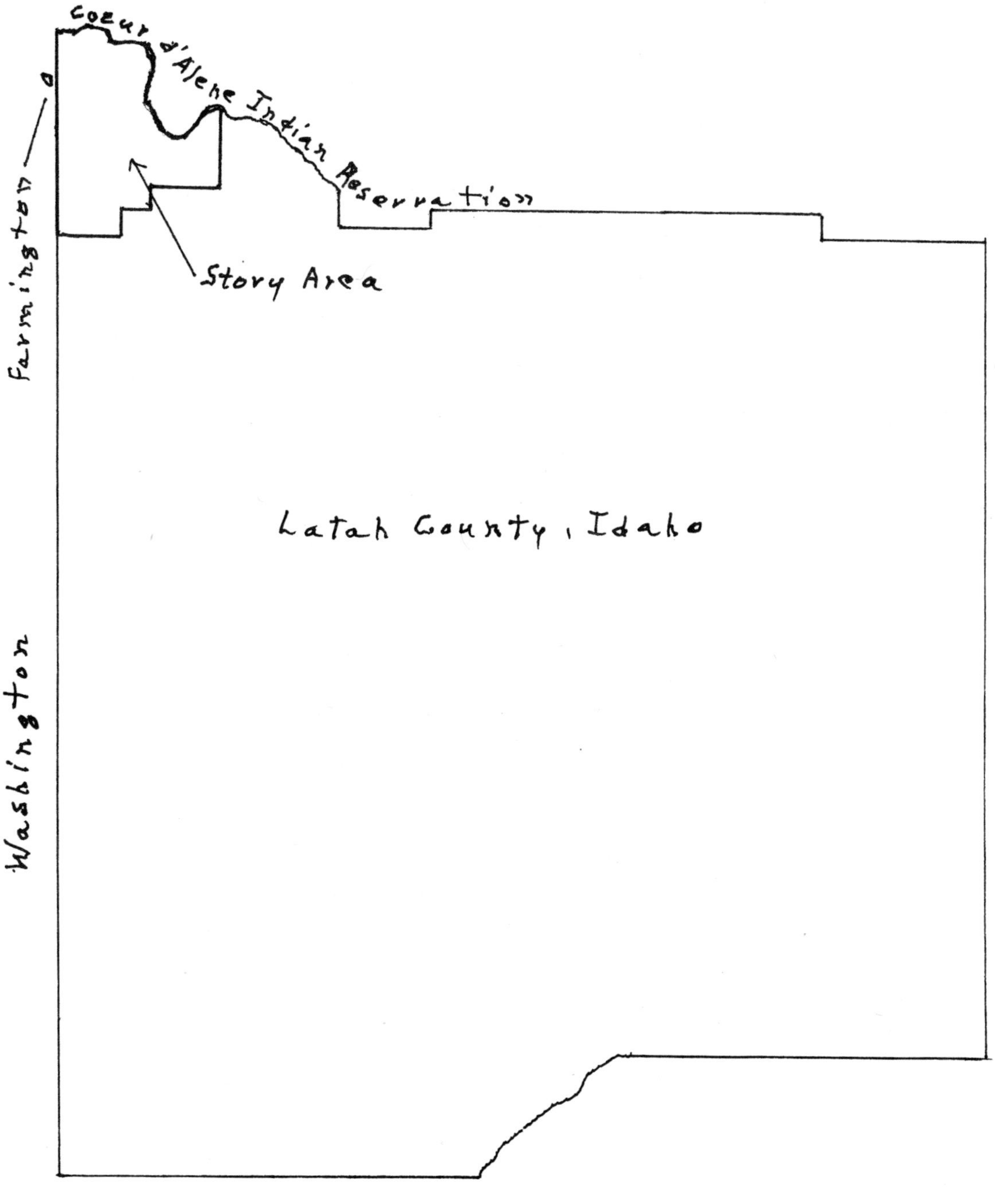

Map No. 3. Latah County, showing story area.

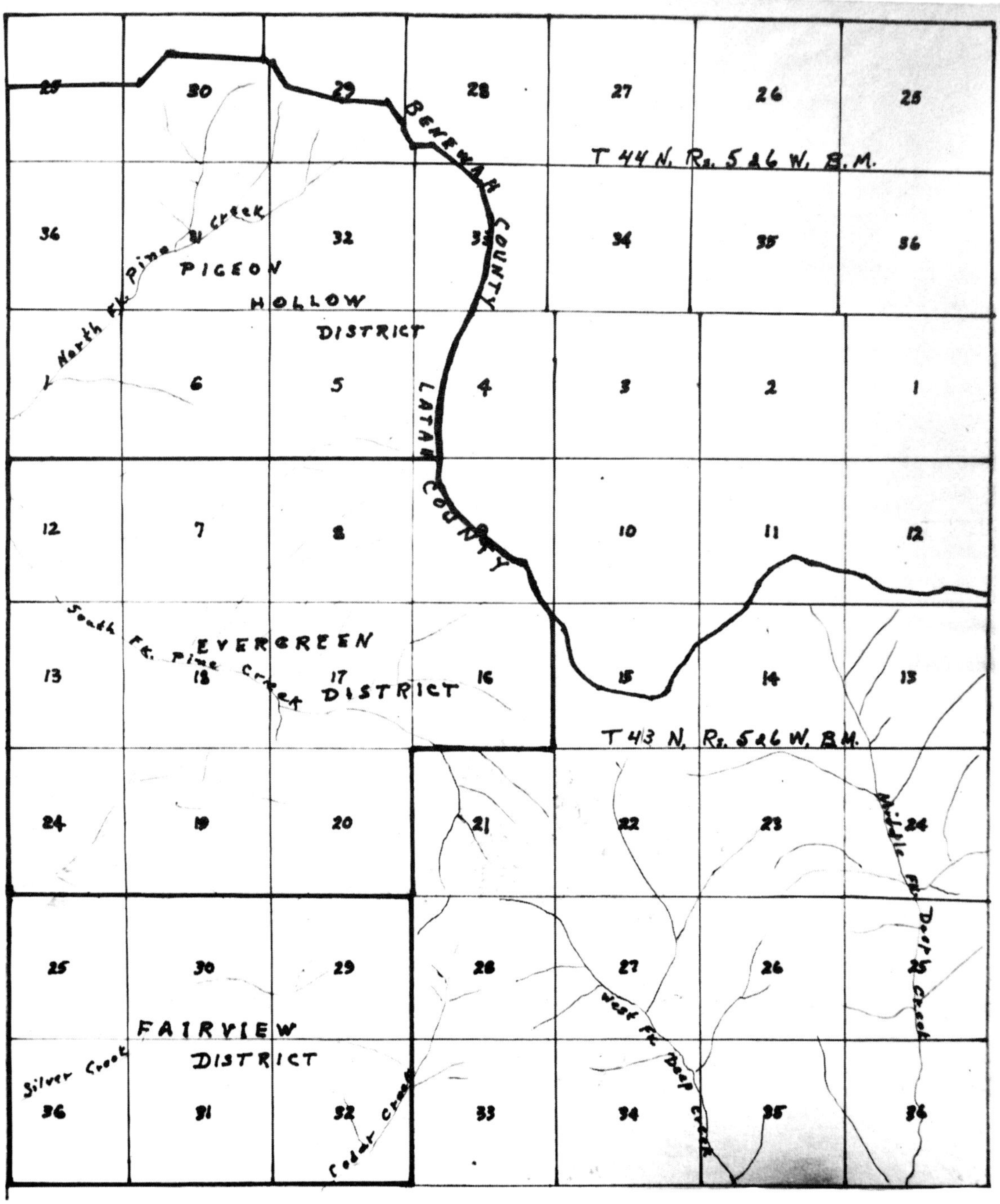

Map No. 4. Story area, showing district boundaries

Fields and Pine Trees

Evergreen—Gem in a Mountain Crown

Evergreen is a very small part of the large state of Idaho. The district has a few square miles of farmland, some farm houses and many Ponderosa pines. Its dimensions are three and one half to four east - west miles from tree line to state line and three miles north to south. Evergreen has a spectacular setting. These mountains are almost entirely covered with trees which spill over the top to border with green the steep fields on the hillsides. The fields slope down and join the round top hills of the Palouse country. The west boundary of the district meets Washington state at Whitman County. The south fork of Pine Creek which drains the land is a tributary of the Palouse River and Evergreen is part of the famous Palouse country.

Geologists say that the soil was blown in on the wind. Inhabitants believe it. Many days of the year a strong steady wind rises and blows from the west all day long. The afternoon brings clouds of dust rolling along and filling the air. Come evening the wind dies down and dust settles showing its presence on the porches and window panes. The process which has been repeating more years than people can count has resulted in a deep layer of soil ideally suitable for farming.

There is evidence that Indians camped in Evergreen. Earl Palmer has found two stone tools such as Indians used for grinding food. The valley was an attractive place for an encampment. Sheltering trees grew along the banks of the creek. The water was clear and cool having just flowed from springs in the mountains. Indians were all around. The nearest tribes were Nez Perce, Cayuse, Palouse, Spokane and Coeur d'Alene. The people of the Coeur d'Alene tribe were friendly and were best known to the early settlers at Pine Creek.

When the white man began to make history in the northwest Idaho Territorial boundaries had been determined by Congress but no lines were yet drawn for the use of homeseekers. There was a vast region mostly covered by grass and largely unexplored. Farmington was not named till 1877 and was not called by its name for another two years. Evergreen was named much later. Pine Creek was a natural feature common to both and gave its name to both areas. In trying to make this story clear all three names, Farmington, Evergreen, Pine Creek will be used wherever it seems best.

The first to come were prospectors. About 1865 they began to work their way north looking for signs of gold and silver. Evergreen was not overlooked in the search. Prospectors dug some holes in the hills but did not find precious metals.

Opal Ross

First Settlers Crowded the Surveyors

A few years later came the homeseekers. They arrived from the south and west having first stopped at Lewiston or Walla Walla. They found a good country. By 1869 a few pioneers reached Pine Creek and began to found a little settlement. One early settler reported that while homes were being built the families lived in a prospector's dugout near Evergreen. Pioneers came west to settle in California or Oregon. They heard about the fabulous Palouse country. When the land was opened for settlement many families pulled up stakes and moved to Idaho or Pine Creek to start again.

The Washington side of the line would not be surveyed for another six years. Evergreen area was surveyed during the last months of 1871. The plot was approved and filed in the Land and Survey office in Boise Dec. 26, 1871. Immediately the lands were thrown open for homesteading. Early settlers in Pine Creek lost no time in staking claims and filing applications. Those who first applied asked for preemptions.

The Preemption Law enacted by Congress in 1841 provided that a citizen could claim any number of acres up to 160 of surveyed public domain upon which land he had built a home and made improvements. The preemption gave the right to buy the land for a minimum price of $1.25 per acre.

Lyman W. Davenport was the first man to settle in Evergreen and claim land by right of living on and improving the place. The date was Jan. 1, 1872. Mr. Davenport entered 160 acres in section 13 not far from and in sight of Pine Creek settlement. A patent was granted to him Oct. 19, 1881. Only a few years later the Davenport land was bought by Andrew J. Hopper and his wife Emma. For many years the Hopper family lived there and participated in the affairs of the community and Farmington.

Patrick Mackey came to Pine Creek with the well known pioneer family of Truax. It was March 1872 that Mr. Mackey made a home in section 18 just east of Lyman Davenport and became the second man to claim Evergreen land. He entered a preemption and received title from the government five years following settlement. The place was 161.94 acres. Pine Creek flows through both the Davenport and Mackey lands. The Davenport property now belongs to Eugene Wagner who lives there. The Mackey farm is owned by Earl Palmer who has a home on the east side.

In April, 1872 Jesse Cash was the third man to settle in Evergreen. He had a wife, Mary and a daughter, Fannie. Jesse is remembered as the first teacher in the first school. He had been injured and was crippled. Mr. Cash staked out a square quarter in section 7. He paid for the land at the rate of $1.25 per acre in four payments of $50.00 each. He did not live to close the deal. After his death the heir, Mary applied for and received title Feb. 1, 1882. Mr. Cash's farm was for many years called the Service place. It is presently owned by Bessie Nolan. No home is there.

Fields and Pine Trees

Each of these first arrivals in good homesteading tradition chose a maximum allowance of land where water was available and wood was nearby. They have not left any descendants that we know of in this part of the country.

As of right now, January, 1974 it has been 102 years since the first settlement in Evergreen. When farming started it grew fast. Allow a few years needed to bring the fields into cultivation and it could be truthfully said, "In all those hundred years, almost every year, year after year the soil has been seeded and good crops have been harvested." The soil in unbelievably rich and durable.

Absalom Price settled near Jesse Cash in 1873. He claimed preemption for 120 acres, an L shaped parcel in section 8. He was granted the patent in 1879. Absalom and his wife Margaret, aged twenty-eight and twenty-one years in 1880 were the youngest of the homesteaders.

1873 was a year of depression. Business reverses in the east caused a panic which spread all across the country. Most people suffered hard times but it is doubtful if Evergreeners were much affected. They were already living hard times. They had to get along with what could be made or produced at home.

They dipped candles, made soap, churned cream, sewed clothes, planted a garden, planted an orchard, cleared land for farming, hauled and chopped wood and carried it into the house, pumped water, took care of animals, made fence. They did all the hundreds of tasks they had to do to survive on frontier farms.

First on the list of needs were a cow, some chickens, and two horses. It was imperative to have horses. An annual trip to Walla Walla was necessary to get the supplies the farmer couldn't make; harness or fixin's for making harness, seed, combs, sugar, coffee, salt, tools, shoes, etc. The farmer could get flour by taking some wheat to the flour mill in Palouse City.

Identity Problem Solved

Pine Creek settlement acquired a post office. This happening was important to Evergreeners who did then and always have since received their mail through the Pine Creek, Farmington office. Mr. Brewer was the first postmaster and his office was set up Mar. 19, 1873. Soon after the door was opened the Postmaster General sent a questionnaire addressed to the postmaster at Pine Creek, Nez Perce County, Idaho asking where the office was located. Mr. Brewer replied that he didn't know because no survey had been made. Mr. Brewer answered other questions; that the nearest prominent river was the "Paloose;" that the nearest creek was Pine; that the nearest office on the route south was 'Paloose' Bridge; that the nearest office north was Hangman's Creek.

In 1878 came another questionnaire from the Postmaster General. By that time the survey had been made. Mr. Brewer knew that his address certainly was not Nez Perce County, Idaho. He could locate his post office according to township, range and section in Whitman County, Washington.[1]

The mail was distributed from Lewiston and was carried by horseback. Riders had a hazardous trail to follow. Mail delivery was very irregular especially in winter. Earth shaking events could happen in other parts of the country and northwesterners would not hear till two or three weeks later. The horseback riders were soon replaced by stage coaches.

Evergreen settlers built a schoolhouse of logs in a northeast corner of land that was soon to be claimed by James Conkling. The site was less than a mile from the Idaho Territorial line. Children from Washington Territory, having no school of their own attended the school as well as Idaho children. At a similar school in Colfax in 1872 the term was three months long beginning in September and ending in February. Some time was lost when weather was very cold and snow was deep. The teacher's salary was $50.00 per month.

Evergreen's first teacher was Jesse Cash. Minnie Allen was the second teacher. Two pupils, the Rees girls, liked to tell many years later that the children were frightened when Indians rode by and stopped to stare in the schoolhouse windows. Two other pupils were Lena Harlow and Pearl Rector. They lived to be very old ladies and passed away only recently. The log schoolhouse was used for a community center and for a church. A minister came from Colfax to hold services.

West of Evergreen Mr. George Truax was creating a town from part of his homestead. He made maps and plans. Mar. 19, 1877 he filed with the county auditor a plat for a town site. An historian[1] who studied the situation has said that Mr. Truax was arbitrary in drawing his lines. He arranged to leave out of his town those he didn't want. Even today some homes that look as though they should be within the town are not. The residents of these homes never vote with the townspeople but have a separate polling place of their own. When the town developed a piped water system the left out homes were left out. Mr. Truax named the town Farmington in honor of his native town of Farmington, Minnesota. He was ready to sell lots.

The years ahead were prosperous years. Farmington became the trading post and much more. In due time the people could boast a flour mill, a fruit packing plant, a round house, long, long warehouses and a large population.

Evergreeners and all the people round about could go to church, send their children to high school, consult a lawyer or a doctor in Farmington. They could market wheat or fruit, surplus cream, eggs or butter in the town. They could buy almost anything from fine china to farm machinery, from baby clothes to caskets in the large merchandise stores. They could go to town to meet with friends for a celebration or a dance. Until transportation to other places became very easy Farmington was our town.

[1]Mr. Charles Blickenderfer

Fields and Pine Trees

There came a day when a driver with a team and wagon hauled a load of sacked wheat down the canyon road to Almota. The sacks were loaded onto a river boat and the boat shoved away from the dock into the current. The first load of Whitman County wheat was on its way down the Snake River to Portland. Another driver hauled wheat to Farmington Landing on Coeur d'Alene Lake. The sacks were carried aboard a freighter which chugged down the lake and up the river to the mining towns of Idaho. This would be the pattern for transporting grain until the coming of the railroad.

Indians were being pushed into a corner and they did not like it. The settlers knew well that if the tribes united their strength they could drive out the intruders or murder them all in their isolated cabins. People were uneasy. All this resentment, and fear and stress led to a panic in 1877. For a few days people huddled together for protection. Nothing happened. The tribes did not unite, did not start a war nor make any trouble. However one white man named John Ritchey was killed near North Fork Pine Creek. Nez Perce Indians were blamed for the murder. Ritchey's body was buried in a small cemetery on the Conkling farm. The story has been told in many northwest history books.

Sometimes a pair of homesteaders filed claims on the same day and proved up on the same day. It may have been coincidental but I think it was planned. A claimant had to ride horseback or drive a horse the long trip to Lewiston to appear in the land office and be sworn. Maybe two homesteader friends or neighbors rode together for companionship. Twin filings were made by Adin Davis and James Conkling who applied for land in section 12 May 1878.

Mr. Conkling's quarter was occupied by the cemetery and the schoolhouse. There is reason to believe that the Conklings did not live there long. The farm was taken over by Elijah Breeding whose name has always been associated with it. Even today people speak of "the Breeding place," "the Breeding barn." Elijah came from Tennessee and was of Irish descent. His wife Josephine was born in Iowa. The Breeding farm belongs now to Mildred Powell.

The eighty acquired by Adin Davis lay close to the Washington line. In 1888 Mr. Davis claimed a timber culture of 6.82 acres which extended his holding to the line. Adin and Mary Jane Davis came to Pine Creek very early from their home at Council Bluffs, Iowa. They had three children, Horace, Sarah, and Mary. Mary became the wife of Albert Leonard. They made their home in Farmington. The Davis property has been kept in the family. Albert Leonard, Jr. is the present owner.

The Timber Culture Act of 1873 which was invoked by Mr. Davis authorized a citizen who kept 40 acres of timber land in good condition to obtain title to up to 160 acres thereof. The minimum tree growing requirement was reduced in 1878 to 10 acres.

Congress enacted a Homestead Act May 20, 1862. Any citizen upon payment of $10.00 might file claim to no more than 160 acres of surveyed public clause was later added. Six months after settlement the claimant could buy the land for $1.25 per acre.

John A. Campbell was the first Evergreen settler to file a homestead. It was November, 1878 when he claimed a 160 acre block in section 17. His deed was granted two years later. Mr. Campbell was born in Kentucky, his wife Mary in Indiana.

Outlaws Stay Brief

At some date after Mr. Campbell and his family moved away from the homestead a family of outlaws took over the ranch. Their name was Masterson. On the south side of the place Pine Creek flows through a magnificent grove of pine trees. It was a genuine hideaway and ideal for the Masterson business of stealing horses. It was said the father and sons stole horses in Montana, drove them through the mountains and forests and sheltered and hid them in the pine grove. After the horses had recuperated from their long trip the Mastersons would fan out over the Palouse country selling the horses to settlers who badly needed them. The Mastersons would then collect more horses and drive them east on a return trip to Montana. The traffic worked both ways. On the frontier laws were not easily enforced. Business was brisk for awhile. The boys liked to ride into town and stir up some excitement. The law finally caught up with them and the father and one son were killed. The rest of the family hurriedly cleared out of Evergreen. I made no effort to verify the account but I believe it to be true. Newcomers and new generations have heard the story and repeated it until it has become a legend.

The Campbell homestead belongs now to Carl and Katie Repp. Son Harold Fisher lives there with his family.

Another legend. This one positively confirmed. A saloon was built in Idaho only a few rods from Farmington. The Evergreen saloon ws named Klondike, indicative of the thinking of the time, 1897. Orville Hayden was shot there and soon died. Officers arrested Charles Marshall who was called "Blackey" and Robert McDonald, nicknamed "Dakota Slim." The two suspects were held in the Colfax jail. A mob broke into the jail, seized Blackey and hung him from a window in the courthouse. Dakota Slim escaped being killed by playing a trick on the mob. Afterwards he was tried and acquitted. Some lurid details are recorded in connection with the hanging but that is part of Whitman County history as is most of the incident. Evergreen provided the scene for the crime.

John Rees and James E. McCabe entered homesteads in 1879. The McCabe claim was an L shaped 120 acres fitting nicely east and south of the Adin Davis farm. The McCabes, aged 53 and 51 on the above date were the oldest of the Evergreen settlers. Mildred Powell lived on the McCabe land and is the owner.

Fields and Pine Trees

John Rees applied for a quarter in section 18 where he was a neighbor to Patrick Mackey. The place contained 3.42 excess acres which was labeled Lot 3. Mr. Rees had been a soldier who served in the Civil War. He and his wife, Ann came from Wisconsin to Oregon to Idaho. They had six children, Benjamin, Elizabeth, Stella, Evan, May and Claudia. The family was well known in Farmington.

John Rees made a trade and moved his family to section 7. His name was associated with the latter place rather than the homestead. Anyone in Evergreen or Farmington would understand what you meant if you said "the Rees place" "the Rees hill." The Rees hill is a trouble maker when snow drifts. Many a traveler has been tied up there and had occasion to learn where he was if he didn't already know. The Rees homestead belongs to Leona Schwartz. The Rees place belongs to Bessie Nolan. Neither place has a home.

Development Rapid

Great changes had taken place in the appearance of Evergreen between the time of the first settlement and ten years later. Good log houses were built and lived in. A stout fence surrounded each occupied acreage and some were rail fences. One of the first improvements a homesteader made was to build a fence around his claim. Enough people lived in the neighborhood to provide a little gathering for social affairs or to talk over problems of interest to all. A school had been established in Farmington and Washington children no longer came to school in Idaho.

The name Price keeps popping up in early history of Pine Creek. The father of the clan was Reuben. He and his wife Sarah crossed the plains in a prairie schooner and landed in Oregon with eleven children, two of whom were born on the way. Two more children were born in Oregon. In 1872 the family moved to the new country called "Paloose" and arrived at Pine Creek.

Two of the sons became homesteaders in Evergreen. One son, J.E. Price lived on the farm later owned by Guy Munger. Another son, Charles and his wife Josephine lived on the Breeding place. Their children were pupils in the log schoolhouse. One of Charles Daughters, Winnie was a lifelong resident of Evergreen.

Absalom and Samuel Price each entered an L shaped quarter in Section 8. These blocks and 120 acres already preempted by Absalom all fitted together like pieces of a jig saw puzzle. David Martin's homestead of 160 acres was the same shape and slipped neatly into the puzzle completing the design. The claims were entered beginning in 1880. John Herrider bought the remaining 40 ancestor. To obtain a parcel of land with your very own name on the deed was no small project. Courage was needed to hang on through all the deprived years.

at public sale in 1890 and all of section 8 was taken. Mr. Herrider later bought 240 acres of the Price land to add to his 40. Herrider's daughter, Fannie and her husband August Danielson lived there. No home is on the place now and the land is owned by August Wagner.

Another pair of homesteaders were Albert Queener and Enoch Rector who applied on the same day in 1880 and proved up on the same day.

Mr. Queener claimed 160 acres in section 24. The Queeners had a lot of livin' on their way to Idaho and took their time getting here. Albert and his wife Mahala were born in Missouri and one son Richard was born there. Two years later Nancy was born in Colorado. Three years later Ida and two years later Sarah and two years later Susan were born in Oregon. Three years later Nora and four years later Myrtle were born in Idaho.

Rector's land was a long, narrow strip, one quarter mile by one mile lying north of the Patrick Mackey place. In his application Mr. Rector declared that he and his family had lived on the land since May 1876. He stated he had built a home, built a barn and a smokehouse and fence surrounding his acres. He said he had 80 acres to wheat, barley, oats and garden.

The land officials asked questions to which Rector answered: that he was a native born citizen of the United States; that he had never before claimed a homestead; that the land had no mineral deposits to his knowledge; that his improvements were worth $500.00 or more; that he was head of a family and thirty-five years old; that the farm was in Nez Perce County, Idaho.

Mr. Rector asked Lyman Davenport, Patrick Mackey, Albert Queener and Adin Davis to prove his claim. All were homesteaders who had settled on land near the Rector home. All have been named in this account. The witnesses verified the statements made by Mr. Rector. A notice was published once a week for five consecutive weeks in the Palouse Gazette, a newspaper printed at Colfax, Washington Territory. The notices caried a legal description of the Rector claim and names of the witnesses. The Publisher of the Gazette sent a sworn statement to the land office in Lewiston to declare the above printing.

Enoch had lived on the place long enough to be a true homesteader and did not have to pay for the quarter. However the parcel contained 2.10 excess acres for which Mr. Rector paid $5.25 or $2.50 per acre. Ora and Enoch had three children, Pearl, Osbert and Ruby. Pearl married Dr. Joseph Leuty who was physician to the people of Farmington for many years. The homestead was later divided into two farms. Leona Schwartz owns and lives on the west eighty. Tracy Ross owns and lives on the east eighty.

Homesteaders Earned Their Land

Nowadays you often hear some person say with great pride, "My father, my grandfather homesteaded that place." He may well be proud to have such an of the contract.

Fields and Pine Trees

It must have seemed unjust to the settler to work so hard for one small quarter or less while the government gave to the railroad company the sections on either side of him and the sections in back of and before him and many other sections in all directions. Who gave Congress the right to give away so much precious land?

Congress granted the Northern Pacific Railroad Company all odd numbered sections of land for 40 miles on each side of its right of way in territories and twenty miles each side of the tracks in states. A strip of land 80 miles wise extended across the country in which every other section was reserved for the railroad.

When the company got around to examine its reserves it found some sections already occupied by settlers who were way ahead of the railroad. Many controversies followed. The railroad lawyers tenaciously clung to the rights of the company and tried to oust the homesteaders from their homes. A case was tried in court and the homesteader won.

After that the government ruled that the railway company could select their lands from an additional ten mile strip on either side of the 80 mile reserve. Even there early comers had already settled and so the government allowed the company to choose any sections. These were called "lieu lands." Sometimes the lieu land selected by the railroad was better than the first choice. If the land was forested it was considered more valuable than agricultural land. When the Northern Pacific representatives came to Evergreen they found that several of the odd sections had one or more occupants in residence there.

The company began to enter its claims in 1885. By the end of the year 1902 the government had given title to 2884.37 acres, a whopping acreage to give away from one small neighborhood. That was not all. The railway company had the right to select 520 acres somewhere else in lieu of the three quarters claimed by Lyman Davenport, Jesse Cash and John Campbell and the 40 staked by Benjamin Boots. Any land which lay on the 80 mile reserve was doubled in value because it would be served by a railroad.

Railroad Building Highly Speculative

The Columbia and Palouse railroad was constructed from Colfax and reached Farmington in 1886. Construction continued through Seltice and onward to Tekoa arriving there in 1888. There were changes of name and jurisdiction in the line as the tracks were laid from one town to another. The rails came into Farmington under the name Columbia and Palouse. From Farmington to Tekoa the line was called Washington and Idaho Railway. Finally the railway became known as the Union Pacific.

The Spokane and Palouse line was built from Spokane to Belmont in 1885 and a branch line extended to Farmington about 1890. The Spokane and

Palouse became a part of the Northern Pacific system. Farmington was served by two major railroad lines.

In section 24 Margaret E. Craig and Jonathan L. Wilson each claimed a quarter that lay in a long south slope down Stratton Butte. Both made entries in 1881. They became neighbors to Albert Queener who was living there. Section 24 is not a full section and a narrow one mile strip was left along the Idaho line. It was 30.42 acres sloping down hill from the top of the Butte to level ground. Frank Wilson homesteaded the land from 1893 to 1898. Two owners, Raymond Simmons and Eugene Wagner hold the four places now. A few parcels and one section of land remained to be claimed.

The last quarter in section 18 went to Christopher C. McCoy who paid for it and was granted a patent April 4, 1884. Part of the land goes up and over Palmer Butte and is forested. Pine Creek flows through a north corner. The present owner is Floyd Palmer who lives there.

Sarah Smith was one of two women homesteaders. She took 141.26 acres along the state line in section 12. Sarah paid cash and received title Aug. 9, 1889. The land is level and Pine Creek flows through and crosses in to Washington. The property belongs to Jack Bennett.

A small place of 40 acres in section 17 was claimed by Benjamin F. Boots in 1883. Mr. Boots served in the Infantry in the Civil War. He came with his wife Sarah and family and nine other families from California by wagon in 1879. One of Mr. Boots' daughters married Monroe Palmer. The Boots' property remained in the hands of the family. The present owners are Jean Hartman, Avis Thonstad, and Marion Ditmore, daughters and nephew to Frances and Monroe.

Early School Site Gift Of Settler

Mr. Quarles was owner of the John Campbell homestead and some other acres in the same section. When it became necessary for Evergreen to have a new school Mr. Quarles was ready with the gift of a tract of land for a school site. The indenture was made Oct. 12, 1887, by and between J.P. Quarles and Margaret Quarles of the County of Nez Perce, Idaho Territory and A.J. Hopper, B.F. Boots and J.P. Quarles, school directors of School District No. 6 to sell a tract of land for the sum of $1.00.

The site was a corner of section 17 at a crossroad. The boundaries were intriguingly defined by stones marked N.W., S.W., and by Pine Creek. Actually there was no mistaking the dimensions of the site. The creek flows in a curve around the corner separating the playground from the field. The crossroads bounds the south and west sides of the plat.

Mr. Quarles declared that the tract must be used for a public school building and be protected for that purpose. If ever the land were abandoned for school purposes the tract would revert to heirs or assignees of the Quarles property.

EVERGREEN SCHOOL — 1907 or 1908

First row	Second Row
Russell Ahrens	Ollie Slavin
Max Federer	Nida Harris
Clarence Felts	Blanche Felts
Roy Harris	Maggie Slavin
Alta Hickman	Ben Davis
Marie Federer	Frank Harris
Fannie Ross	Basil Hickman
Norma Ross	Tracy Ross

Miss Edna Redmond—Teacher

Picture by courtesy of Tracy Ross

Opal Ross

Evergreen school was built. It was plain wood frame style so popular for early day schools. There were three tall windows on each side of the building and a door on the south end. The school was backed by pine trees and a rail fence. Pictures of old school houses show them all looking alike, the plain utility type rectangular, one room building usually having a background of trees.

Inside a row of hooks was fastened to the entrance wall for hanging coats. A big wood gobbling stove heated the room. The children carried pails of drinking water from the home of the nearest neighbor. Their route led through an orchard and the children stopped to sample the cherries or apples whatever fruit happened to be ripe. A well and pump were added to the school some time later. One teacher taught the school whether there were four or forty children, one or eight grades. The building was in use for a school till 1934. From that time a bus or a car took children to Farmington or Oakesdale.

Evergreen School was a place to have good times. Four sets or even more on a crowded evening could take their places two by two for a square dance. Sometimes there were "Literaries." Those who provided the entertainment could often remember for years the "pieces" they memorized to recite at the program. Christmas eve was a bright occasion. Neighbors all came if they could to marvel at the big tree and hear the organ music and the children's plays.

Evergreen school was a place for meetings. The rural telephone company met there for the annual stockholder's meeting. The board sat there when Idaho had an election. That was the day, a long day. Each board member had with her enough food for two meals. Residents from the Pigeon Hollow district north and Fairview district south also voted in Evergreen. It was a time to cast your ballot and also to visit.

The school house is not there now. Honda trails weave in and out among the trees of the playground and along the bank of the creek.

When Idaho was a territory public schools were supported by taxes on certain lands, by fines from breach of penal codes, percentages of fees on toll roads, bridges and ferries. When Idaho became a state in 1890 the federal government transferred the lands which had been reserved for schools. Every 16th and 36th section of every township was school land. The land was worth $10.00 an acre at that time. Only part of it could be sold each year. In Evergreen, section 16 school land was timber. Part of the section was soon sold in small lots and the trees were cleared to permit farming.

For a long time the roots and stumps left in the ground were a hazard for a man working with a horse drawn plow. Says one old timer, "You might be plowing with a two bottom plow and five horses. You could ride down hill but if you had any respect for the horses you walked uphill. If the plow struck a buried stump any one of several things could happen. The plow beams might twist out of shape; you might ram your chest against the lever; the plow might throw out and throw you off; the double trees might break and the horses be freed. It the horses had been plowing a few days they were happy to stop where they were." Part of section 16 is now within McCroskey Park.

Political Progress, But Economic Disaster

The first Idaho Territorial Legislature attached all the land north of Nez Perce County to Nez Perce County. In May 1888 the United States Congress created Latah County from part of the land. The county seat is Moscow. Ten election districts were named of which Pine Creek, now Evergreen, was one.

People shivered through the bitterly cold winter of 1889-90. Subzero temperatures were the rule everywhere. The hardy winter wheat was not damaged and when spring came the Palouse country looked good. In 1890 E.C. Warner editor of the Colfax Commoner wrote: "Less than ten years ago the editor of the Commoner drove across the beautiful rolling prairie lands of Whitman County and less than ten days ago made a similar trip. Upon our first trip the thousands of acres of undulating prairie lay unbroken and clothed only in a luxuriant growth of bunch grass. Here and there could be seen a few rude farm houses and small fertile tracts in cultivation while the wide rolling hills were used as pasture by owners of cattle, horses and sheep. Herds of stock roamed at will. The view disclosed by the last trip was widely different. The uplands and fertile flats have become continuous blocks of farms. Good roads, excellent bridges run in every direction. Neat barns and substantial houses were seen on all sides."

A convention of territorial delegates met at Boise City to frame a constitution for statehood.They presented it to Congress and Idaho was accepted into the Union as the 43rd state. The date was July 3, 1890.

There was time of prosperity, then in 1893 disaster struck. An old timer recalled, "It started to rain on the 28th day of August and she just kept on raining for four months, steady. The sky never seemed to clear and the sun didn't shine. Everything that was planted either rotted where it stood or became so water logged it couldn't properly develop. In November the snow began. It grew colder and continued to snow and the ground froze solid. There were no places to spread out and dry the small lots of wheat that were saved. Snow covered all natural feed and stock died of starvation. Panic and depression made hard times worse. It was a tragic year. The country was a long time recovering."

Fruit Important Early Product

Farmers planted orchards particularly apple and cherry trees. The apples they produced lacked the aristocratic, polished appearance of today's Delicious and Winesaps but they are precious in memory. Anyone who ever took a bite from a Red June apple would not forget how good it was. The orchards had

Gravensteins which were unrivaled for making pies; crabapples, rich in pectin so necessary to firm up jelly; Rome Beauties and other varieties. The Beauties were winter keepers and hung on the trees till October and were picked one jump ahead of Jack Frost. Some of the early varieties were still grown today and others are hard to find.

Early settlers could buy fruit trees. The first edition of the Palouse Gazette appeared September 29, 1877. On the front page was an advertisement offering fruit trees, berry bushes and flowers for sale. The nurseryman stated the young trees were started and acclimated in the Palouse valley ten miles below Colfax. Besides apple trees he offered cherries, plums, apricots, prunes, and pears. He offered gooseberries, raspberries, and other berry bushes, roses, snowballs and clematis.

Harry Davis, a homesteader planted two orchards on his land on the hill. He planted the above named kinds of fruit. In every part of the season some fruit was ripening. Twice a week from summer to late fall Mr. Davis loaded his hack with fruit and drove the long way to Potlatch to peddle the fruit on the street.

George Godwin had a different type of fruit farm. He bought 40 acres of school land and used some of it for a large planting of strawberries. He lived in Farmington and commuted to his work of tending the strawberry patch. For Mr. Godwin commuting meant walking five miles to work and five miles home.

A few years before and some years after the start of the new century fruit was an income crop in Evergreen. Many boxes were packed in Farmington and shipped from there. Some of the old trees still survive overgrown with brush and surrounded by other trees. One old apple tree bears good apples having no worms. The last few years it has been pruned and cared for by way of reward and encouragement. An old apricot still lives. Its fruit is ready to eat at harvest time. What fun to jump off the combine, walk a little way into the brush and fill a sweaty hat with apricots!

Section 20 is steep north and west hillside and forested mountain tops. It was last to be settled. Five men claimed the land. Thomas B. Queener, Frank Lewis and William H. Davis homesteaded 160 acres each. Frank Reid claimed 40 acres. Peter McKenzie invoked the Timber and Stone Act for 120 acres. The law enacted in 1878 provided for the sale of timber and stone lands unfit for cultivation at $2.50 an acre up to 160 acres.

Frank Reid and Peter McKenzie sold out and departed early leaving no footprints in Evergreen. Tom Queener and his wife Josephine had a large family. The children attended the Evergreen school. Frank Lewis kept his land. His cabin is the only homestead cabin left in Evergreen. The logs are handsomely fitted together. The workmanship shows a great deal of skill in the use of an axe. Lewis was a Belgian of some worldly experience before he came to remote Evergreen to take a homestead. He and his wife moved to Farmington and are remembered by a few of the people there.

Corner of the Frank Lewis cabin, now owned by Earl Palmer. The homestead was staked in 1883 by Lewis. Note the way the logs are morticed.

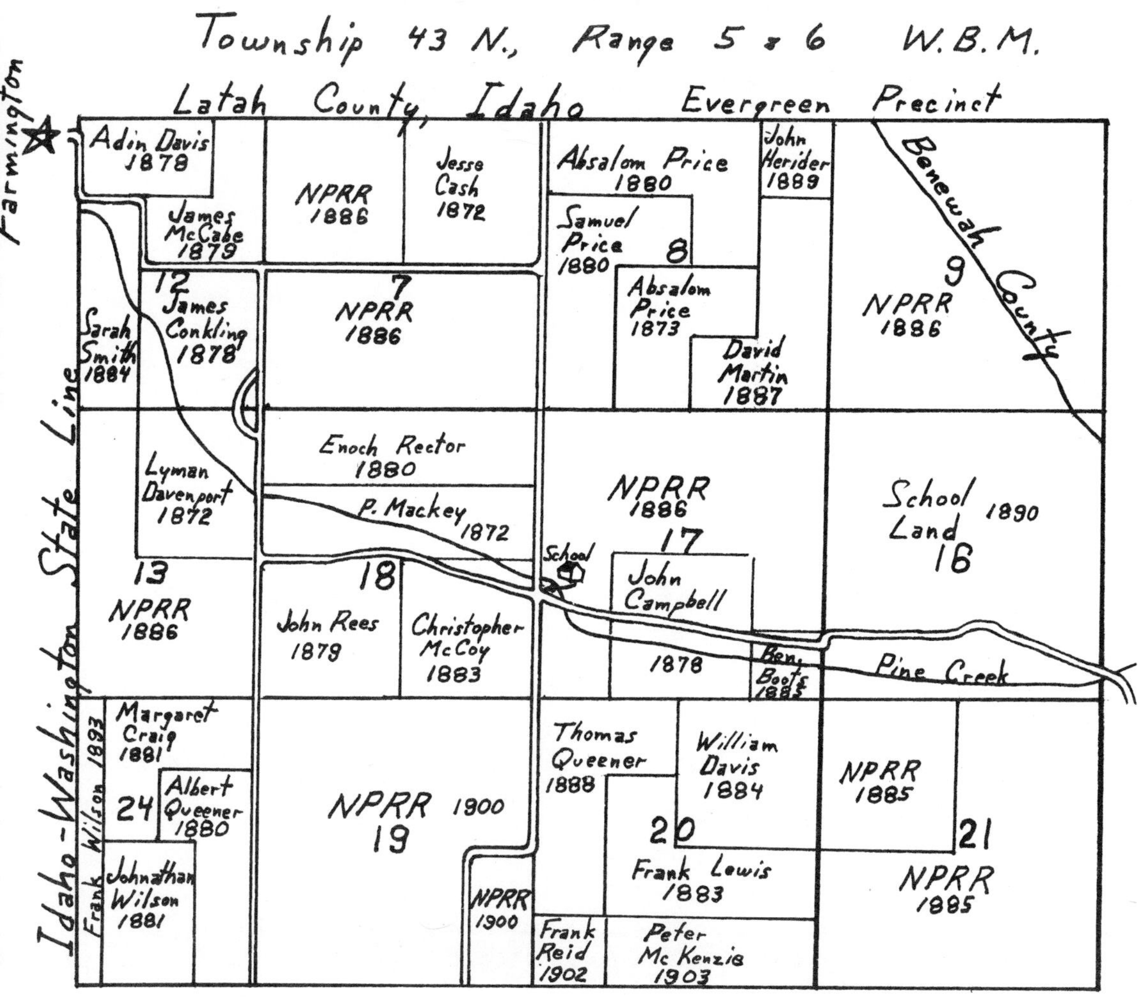

Homestead Map 1872 – 1904

W.H. Davis and his wife Josephine (who had the orchards) stayed on and made their presence felt in the neighborhood. Harry was a fiddler who played for programs and dances. Josephine accompanied him by chording on the organ. Their son Benjamin served with the army in France in World War I.

The land of section 20 was claimed and won between 1888 and 1904. Earl Palmer, Fannie Ross, Bessie Nolan and Potlatch Forests are the owners now.

Many Changes, Many Departures

Some statistics about Evergreen homesteaders: Twenty-five were named in this account. We know the ages and origins of twelve and their wives. They came from many states. Missouri contributed the largest number and Indiana was first runner-up. All were native Americans except one wife who came from Wales and Frank Lewis. The average age was thirty-eight.

In 1914 only five parcels of Evergreen land still carried the name of the homesteader, Adin Davis, Herrider, Boots, Rees and W.H. Davis. Now in 1974 only two places belong to descendants of first owners, Benjamin Boots and Adin Davis.

The railroad property was sold in generally small acreages. People bought and moved in. Evergreen population was large in the 1890s and early 1900s. The phrases so often repeated in this account, "no one lives there", "no home is on the place," these words describe what has happened in Evergreen. It has gone the way of all farming communities. There are nine homes and one house which is rented part of the time. Two other homes are not in the district but the families living there are Evergreeners anyway. By 1904 homesteading days were over in Evergreen. All the available land was staked, filed, fenced, cultivated and otherwise improved. It was a heap o' livin' in thirty-two years.

In every locality the same events occurred at about the same time. Everywhere in the Palouse Country schools were established, post offices and mail routes organized, rails were laid, small towns ten to twelve miles apart came into being. Almost every town had its big fire and some towns had more than one fire. Thin wooden buildings were destroyed and were soon replaced with better, more substantial buildings. Life moved so fast there was little time for dismay or regret.

When Congress enacted the free land laws it desired the country to be settled. It wanted railway lines to span the continent and to reach every whistle stop where something could be delivered or picked up. It happened. All came true with a headlong rush.

First Telephone Cooperative; Rules Important

Residents of Pine Creek, Idaho met at the Evergreen Schoolhouse April 8, 1907 to organize a phone company. Nine were present. J. B. Johnson was elected president, Milo Palmer, Vice President, George Boyer, Donald Hopper and Park Harlow, directors, S.A. Ross, Secretary and Treasurer. The Evergreen Rural Telephone Association was formed to build a phone line and connect with a central exchange at Farmington.

The members agreed that each line right would be worth $20.00 and that no more than twenty line rights would be sold. The members agreed that each one could furnish poles and do what work he wished. The price allowed for poles would be 30 cents each. Man hire would be $2.00 per day or 20 cents per hour. One team, one day would get the owner $2.50. The company ordered ten ringer phones at $12.75 each with a discount of 50 cents per phone and 2 per cent off for cash.

The Constitution laid down some firm rules of conduct. So stated: The board of trustees shall have the right to impose reasonable fines on all persons using the line under the following conditions.

1. For impersonating another individual.
2. For the use of profane or obscene language.
3. For intentionally interrupting or listening at the phone when others are using the line.
4. For using the line an unreasonable length of time.

No one minded the rules. Rings were like three shorts, two longs and one short, two longs and two shorts, etc. Every ring could be heard by every household on the line. If a person wanted to know what his neighbors were talking about he "rubbered."

As soon as spring work was done the phone line was built. Poles generally followed the fence line. Dues were $2.00 per year or more if assessments were made for maintenance. If a family moved away they could sell their phone right to the company or to another family or person. At one time there were nineteen on the line. The telephone exchange was at Paddock's store and thirteen other rural lines met there. Later all the lines cooperated to set up a central office with long distance convenience.

The Evergreen Rural Telephone Association met annually or when the President called a meeting. T.J. Torpey was President and S.A. Ross Secretary for many years. The rural line was in operation and adequate for more than half a century. The last meeting of the Association was held in April 1958.

In 1956 General Telephone Company purchased part of a lot in Farmington for the site of a telephone office. The Company bought the fourteen rural lines and rebuilt them to be compatible with new dialing equipment. In October 1957 the Farmington area was cut into the dial system.

Opal Ross

Rural Roads, RFD, Electricity

Idaho became better looking when the state set into motion a program for improving farm-to-market roads. One purpose of the project was to provide work for farmers who were badly depressed by the low price of wheat in that year of 1934. A self appointed committee of Evergreen farmers went to Moscow to meet the Governor and ask for their share. They got a promise. A crusher was placed in a rock quarry on the Felts place to grind the rock for surfacing. Farmers hired out their horses to do the work and many farmers drove their own teams. The old road bed was widened, graded, smoothed. Fences were set back, curves softened and hills reduced. A layer of crushed rock was spread over all to take the place of the mud. This wonderful improvement came none too soon. People were needing to drive their cars in all kinds of weather.

In the spring of 1935 routes #1 and #3 from the Farmington Post Office were consolidated. Harry Duncan was assigned to the double route and began to bring mail to Evergreen. He kept a sled and wagon and horses to use during snowy or muddy weather. The morning train brought the mail about seven o'clock and the carriers left the post office soon after. When driving a horse drawn vehicle the day was long and hard and lasted till dark. Harry remembers keeping his lunch near his foot warmer to prevent it from freezing. This was not always successful and he often had to eat frozen sandwiches.

Mr. Duncan's words, "have memories of roads blocked with snow. I remember one winter when I traveled across country cutting twenty-one fences and traveling on top of snow crust. No one seemed to object to having his fence cut. They were only too glad to get their mail. In one place I recall where the snow was so deep I could reach out and touch the telephone wire."

Carrying the mail on a rural route wasn't all bad. Sometimes the carrier found a gift for himself in a mailbox or a patron would bring him hot coffee. He made many friendships on the route. When Mr. Duncan retired in 1957 his patrons gave him and his wife a party. After Mr. Duncan's retirement George McNair took his place. He carried the mail for many years and had problems fighting drifting snow.

In Evergreen power is provided by a rural electric called Clearwater Power Company. The company is a cooperative and is owned by consumers. This service became possible when Congress passed the R.E.A. Act in 1936. Two years later a contractor set the poles and strung the wires in Evergreen. The power was first turned on in November, 1938.

Electricity was considered a luxury and some farmers wondered if they could afford what it would cost. Now, the company says and we all say, "Electric power is a vital necessity of day to day living." "It has improved the quality of life."

Fields and Pine Trees

Forested Parklands Abound

The mountain tops adjoining Evergreen are home for a state park which was created by Virgil McCroskey. Virgil bought the land a few acres at a time as he could persuade the owners to sell. He cleared away just enough timber to make narrow roads providing access to every area of the park. All together these roads comprise the Skyline Drive. The park is a wonderful natural woodland and is meant to be kept that way. There are some picnic tables and a fireplace but few other improvements. Birds, deer, chipmunks and other wild animals live there. Sometimes bears and elk wander into the park. There are wildflowers, syringa shrubs and huckleberry bushes. In a good year the supply of purple huckleberries will equal the demand. Tamarack and cedar and other trees grow in the park but the Ponderosa Pine is the most distinctive tree to be found. People are permitted to gather waste wood but are not allowed to cut good trees.

Mr. McCroskey dedicated his park to the memory of his mother, Mary Minerva McCroskey. He gave the park to the State of Idaho bequeathing in his will money for maintenance. McCroskey Park is used for recreation and many people go there. Drivers are sometimes startled to see a car approaching when the road is only wide enough for one car. Both drivers calculate and maneuver, one driver sometimes backs up but they always seem to get by safely. A day in the woods to hike or picnic is available to anyone. To Evergreeners this is a pleasure only a few minutes from home. In the last few years a new recreational use has been found for the Park. Snomobilers find the roads to be ideal for their exciting sport.

Camp Virgil McCroskey is a boy scout camp adjacent to McCroskey State Park. It is open to all scouts in the Inland Empire Council. Boys camp there is good weather or very unpleasant weather.

For a further background Evergreen has the vast St. Joe National Forest. The preserve extends east in a broken line to the St. Maries River and the Boise Meridian. this is only one portion of the forest. Another great section begins a few miles farther east and reaches to the Bitterroot Mountains and the Montana border. The western boundary is the north-south main road through Evergreen. The St. Joe National Forest is something about which to be proud. There stand some of the finest white pines in the whole world. The white pine is one of the greatest treasures of Idaho. Timber sales and grazing fees provide a revenue from the forest. 25 per cent of the income within the county is used for schools and roads.

Northeast over the hill in Benewah County is the Coeur d'Alene Indian Reservation. On the domain at DeSmet are a mission and a school. Mission of the Sacred Heart was founded by Father Peter DeSmet in 1842 at another place in Idaho. It was moved and reestablished at DeSmet in 1876. Jesuit Fathers administer the mission and they have a most endearing motto, "Among the

Coeur d'Alenes." The school is independent of the Jesuit Fathers and is in charge of the Sisters of Charity of Providence. It is called "Convent of Mary Immaculate." Seventy-two children go to school there now, one half are non-Indian.

The Larger Scene

East over the hill is Idaho's famed North South Highway 95. The road was constructed fifty years ago to satisfy a desperate need for better communication between the two parts of the state. 95 is as long as Idaho. North it goes over the border into Canada. South it crosses the state line into Oregon. It is not adequate for present day traffic but is still a beautiful, scenic highway.

Over the hill and southeast is Potlatch, an Idaho town of importance. The Potlatch Lumber Company built the first mill there in 1906. The lumber mills are the dominating force in the town. Evergreen's road district and school district are centered in Potlatch.

This side and that side of the line: Evergreeners find some difficulties and some advantages being residents of Idaho and belonging to Washington. When someone says to Mr. Evergreen, "What is your telephone number?" he is thrown into a quandary about what to answer. The area code assigned to him in 509 which is the code number for eastern Washington. Yet people calling him have to give the Idaho code number. He stammers and finally gives both code numbers and lamely tries to explain why he has two area code numbers. However when Mr. Evergreen places a call which must inevitably go through Washington from Idaho he has the advantage of being on the other side of the line and gets a cheaper rate for his call.

There is the problem Mr. Evergreen faces when he presents a driver's license for identification. There he is with an Idaho license and Farmington, Washington address. He usually has to explain. In nearby towns and even in Spokane clerks have generally learned to understand but in Seattle and some other places they don't get it. Sometimes they are inclined to be suspicious.

The tax situation has become so tight that a buyer can't beat it. For a long while Mr. Evergreen could buy in Washington, have his purchases delivered to his home in Idaho and escape paying the sales tax. No more. Now he has to pay the sales tax wherever he buys and must pay the Idaho income tax as well. However if he buys a large item like a car he is entitled to pay the smaller tax because he lives in Idaho.

Many roads cross the Panhandle west border where the traffic flows from one state to the other. It is said that a traffic violator or criminal can flee over the line with the law close on his heels and escape. Jurisdiction stops at the line. Not true in Evergreen. For five months of the year Evergreen is a dead end. There is no place to go except back into Washington. And so this small part of Idaho is quiet. Very few wrongdoers streak through.

Today's Families In Evergreen: February, 1976

Raymond and his daughter Rise are the third and fourth generations of the Simmons family to live in Evergreen. At the present time Raymond owns and operates the farm. He and his wife Thelma have a home in the southwest corner of the district. Nearby lives Rise and her husband Gary Parberry who shares in the farm work.

Eugene Wagner and his wife Laurene live on the west side. Their children--Karen, Connie, Jack, Janice, Anne, Kevin, Leon, Lynn, Gretchen. All are married or away at school.

Robert and Betty Wagner. Robert's youngest son, Jeff is at their home part of the time. Gary, Russell, and Rob are married. Rick passed away.

Anne and Ronald Miner and their small children live farthest east in Evergreen. Ann is a daughter to Eugene Wagner.

Two brothers, Floyd and Earl Palmer are close neighbors living near the center of the district. Floyd married Ruth Wright and their children are Greta, Carol, Alvin. Earl married Hazel Johnson and their children are Lois and Larry. The children from both families are grown and living elsewhere.

Harold and Carol Fisher live on the old Campbell homestead which has probably seen more history than any other ranch in Evergreen. Randy and Karen are away from home. Son Roger is the only child of school age in the district.

Add Tracy Ross and his wife Opal and the list of present day Evergreeners is complete. Their children are Anita and David. Tracy has the longest continuous resident in the district.

A Few of the Many Former Families of Evergreen:

The Simmons, George and Helen bought a homestead right in Evergreen in 1879. "Indians going north or south traveled by their house regularly and in great numbers. It was the main trail between the Nez Perce and the Coeur d'Alenes."[1] Three children were victims of typhoid fever. Frank, Chester and Ross lived to grow up. Chester was the son who farmed the home place and acquired more land. He married Emma Weible. Raymond is their son.

The Hopper family came to Evergreen in 1884 and settled on a ranch near

[1]Told to Raymond by his grandfather George Simmons.

Norma, Tracy, Fanny Ross.

the Washington Territorial line. There were three children, Anna, Rex, Donald. Life was not easy for Emma Louise (Mrs. Hopper). Her younger son Donald was stricken with polio and crippled for life. Her husband, Andrew died when the children were young. Donald grew up and farmed the land and lived there with his wife Bessie. After their death Rex and Rea took over the home place. Their children--Jack, Emma, Marilyn. Jack was killed overseas while serving in the army.

In 1897 Milo and Catherine Palmer bought land in Evergreen. Four of their sons established homes in the district. Three of the sons, Monroe, Floyd, Earl have been accounted for in this story. The fourth son was Roy. Roy and his wife Bernice had five children--Letha, Melvin, Florence, Velma, Doris.

The Boyers came from Indiana in 1901. They were six, Grandfather Boyer, George and Rose and their children, Bernice, Guy, and Cletus. Another daughter, Doris was born in Idaho. The Grandfather bought a quarter in section 7 in Evergreen. On the place was a one room log cabin with a single two by two foot window. George later rebuilt the house. The Boyer land was sold to the August Wagner family. Bernice and Guy have passed away.

Archie (S.A.) Ross, his wife, Jeanne and their children came from South Dakota to Evergreen in 1904 when homesteading days in the district were just over. Other members of the family were already living in nearby places in eastern Washington and Idaho. The Rosses bought land which had been homesteaded by Enoch Rector. The family moved into a small house on the place. In a few years they were joined by Sophia McNair who was an aunt of Jeanne's. The Ross children--Tracy, Norma, Fannie. Tracy is the present owner of the land. His children--Anita and David.

Until very recently Mildred Powell was a resident of Evergreen. For the last half year she has been in the Colfax hospital unable to come home. Mildred is the daughter of Frank and Sadie Leonard who at a very early date bought a place in the district not far from the state line. Mildred married Plutarch Powell.

Henry and Sarah Brunner came to Idaho about 1909. Their children--Ben, Essie, Archie, Winnie. The Brunners bought the old Rees homestead. After Henry's death Ben farmed the land. He and his wife Elma had one son, Delbert.

A more detailed history of each of these families can be found in the McConnell Museum in Moscow, Idaho.

Isaac and Leona Schwartz, two sons, Carlyle and Mike, three daughters, Marsha, Marjorie and Judy. The Schwartz family lived on the west side of the old Rector homestead and were long time residents of Evergreen.

Two other families lived for awhile in the district and made lasting friendships here. Wallace and June Hodges and their family, Alvin, Byron, Vicki, Sharon. Emmett and Julia Howard and their sons, Wayne, Bruce, Donald.

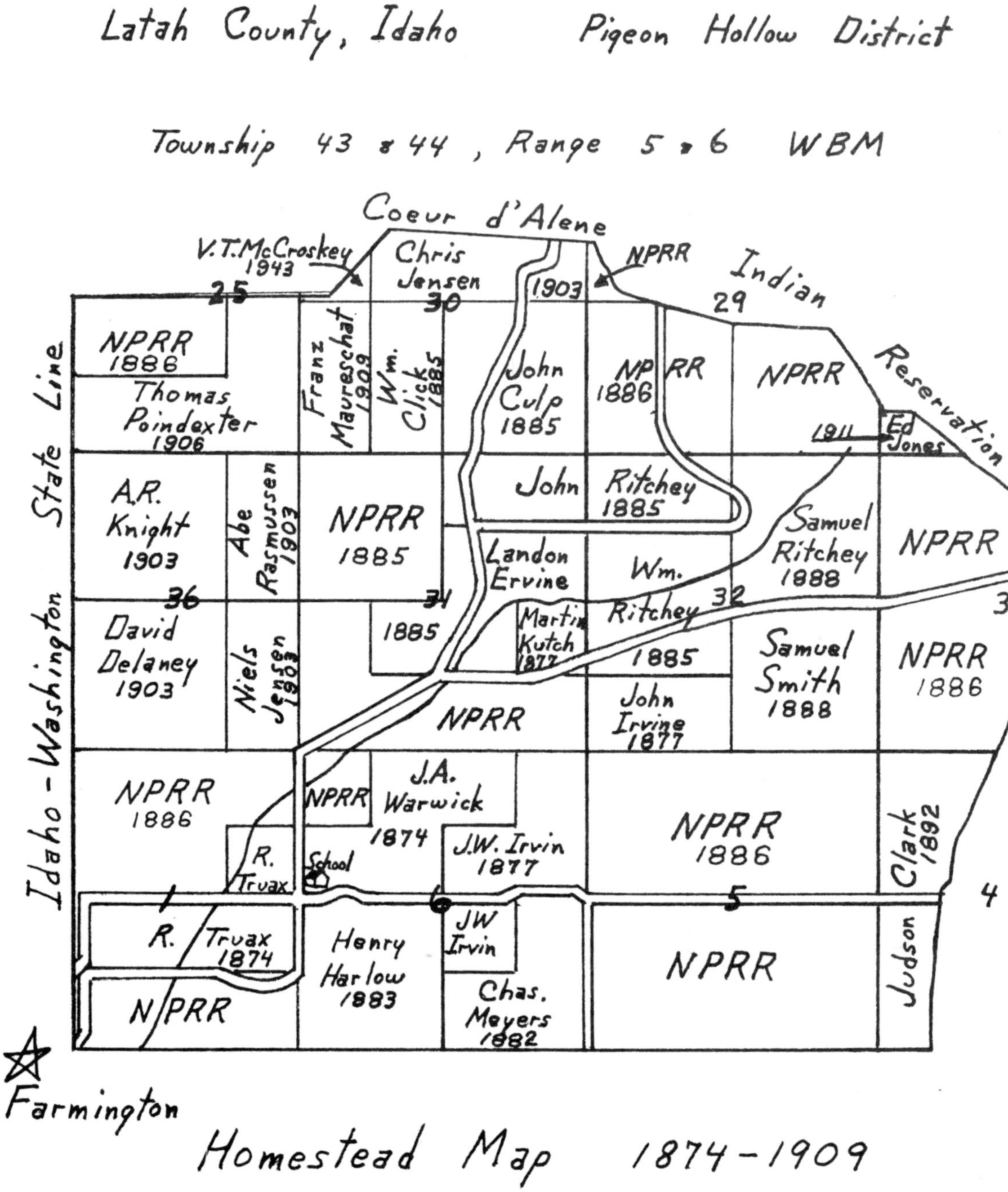
Latah County, Idaho Pigeon Hollow District
Township 43 & 44, Range 5 & 6 WBM
Coeur d'Alene Indian Reservation
V.T. McCroskey 1943
Chris Jensen
1903
NPRR
25
30
29
NPRR 1886
Thomas Poindexter 1906
Franz Maureschat 1909
Wm. Click 1885
John Culp 1885
NPRR 1886
NPRR
1911
Ed Jones
State Line
A.R. Knight 1903
Abe Rasmussen 1903
NPRR 1885
John Ritchey 1885
Samuel Ritchey 1888
NPRR
Landon Ervine
Wm. Ritchey 1885
36
34
32
33
David Delaney 1903
Niels Jensen 1903
1885
Martin Kutch 1877
Samuel Smith 1888
NPRR 1886
NPRR
John Irvine 1877
Idaho-Washington
NPRR 1886
NPRR
J.A. Warwick 1874
R. Truax
School
J.W. Irvin 1877
NPRR 1886
Clark 1892
1
6
5
4
R. Truax 1874
Henry Harlow 1883
JW Irvin
NPRR
Judson
N/PRR
Chas. Meyers 1882
Farmington
Homestead Map 1874-1909

Fields and Pine Trees

Pigeon Hollow—Indian Pathway

A look at the map shows Pigeon Hollow perched like a Santa Claus cap atop Latah County. No one remembers how this small area of farm land and trees acquired the name "Pigeon Hollow" except the obvious reason that many pigeons lived in the barns thereabouts. The name was probably bestowed when in 1886 an imaginary line was drawn east to west across the county (then Nez Perce County) to designate the border of a school district.

A spur of mountains from the Coeur d'Alene Range curves around the north and east sides of the valley. The irregular summit of the mountains is the county boundary line and is also the divide between Palouse River drainage and Hangman Creek drainage.

During the two years from July 1867 to July 1869 Allen M. Thompson surveyed two tiers of townships along the western border of the Territory of Idaho which survey included Pigeon Hollow. As he worked his way north he made notes in his field book about the type of soil, the terrain and the direction of Indian trails.

Indians made well defined trails in their many journeys. Sometimes by choice they went "over" and not "around". The higher ground provided dryer footing as well as advantages in seeing what was going on. The trail had to be chosen to divide into daily lots from one spring to another spring or creek. After the best route was found the Indians traveled the same path over and over again. All the pounding of mocassined feet and cayuse hoofs deepened the trail. Pioneer explorers often followed these thoroughfares finding them the best way to go.

One such trail traversed Idaho from Lapwai north, crossing the Clearwater River near where it flows into the Snake River. The track roughly and sometimes closely paralleled the western territorial border before the border was ever determined. The trail crossed Evergreen east of Queener's Butte, now called Stratton Butte.

The surveyor made very plain the direction of the trail in Pigeon Hollow. He wrote in his notes, "Between ranges 5 and 6 W., T. 44 N., between sections 31 and 36 Indian trail bears southwest and northeast. Sandy hill slopes to the southwest." It was an excellent trail. A long, easy walking, easy riding slope took the traveler from the valley floor to the summit of the mountains. Traffic over the trail was heavy and gave to Pigeon Hollow a rich and colorful history long before white men invaded the land.

Nez Perce Indians were the most frequent travelers over the trail. It was one of their routes for going to Montana to hunt buffalo. They traveled to Lake Pend Oreille, thence eastward through the mountains along the banks of the

Clark Fork River passing the Coeur d'Alenes on their way. Father DeSmet in his long black robe rode over the trail. Nez Perce often went north to visit the Sacred Heart Mission. We suspect that both the Coeur d'Alene and Nez Perce people traveled north or south just to get together for fun and games.

In the fall of 1873 another survey was made in greater detail. Thompson Meldrum surveyed and subdivided townships 43 and 44 N. R. 5 and 6 W. In his field notes he left a record of his method. He wrote, "Set a low post four feet long four inches down for one quarter section corner in mound of earth two feet high with charred stake, pits, etc."

Sometimes the surveyor used a large stone for a section corner marker and if he were in the woods he blazed a tree. Mr. Meldrum added a comment to his work notes. "Land rolling. Soil first rate." The plat was approved by the Surveyor General Jan. 1, 1874 and Pigeon Hollow was opened to homesteaders.

First Come, First Choice

Oct. 7 Mr. Richard Truax filed a preemption in the newly surveyed district in Nez Perce County, Idaho. He claimed 160 acres of choice Palouse farmland. His claim bordered the west territorial boundary and was in sight of the flourishing settlement that would soon become Farmington, Washington. Water and wood were abundant and close at hand. The soil was "first rate." The land was level. It was a good homestead. The best! And why not? He was the first one there. Two years before he had built a rude little cabin and he and his wife Sarah and daughters, Clara and Bertha moved into it.

Sarah was born of Holland Dutch and Wales parents.

Mr. Truax paid $1.25 per acre for the farm. He went to Lewiston to prove up. George O. Briggs and James F. Conkling went with him to witness that he had built a home and lived in it; that he had plowed, fenced and cultivated over twenty acres of land and that these improvements were over the value of $100.00.

Life was hard. Almost everything the family needed was scarce or not available at all. In a few years Mr. Truax left the farm and followed other occupations in other places. But he and his family returned and his descendants unto the fourth generation lived in Farmington.

John Warwick filed his preemption only days after the Truax claim. Because of a discrepancy in the survey line Mr. Warwick got 163.85 acres. His farm was also a good farm.

The homesteading of the area would continue for thirty-seven years and the last patent would not be granted till 1917. The size of the claims varied from 19 to 173 acres. There were conflicts with the Northern Pacific Railroad. Sometimes the settler arrived early and was firmly in place before the railroad company representative entered his right. The railroad would then take parcels of land here and there "in lieu of."

Fields and Pine Trees

Pioneers often settled and lived on the land for awhile and made improvements and filed or did not file claims. Often they abandoned their places and sold their rights. Others carried on and secured the patents.

The pioneers who came to Pigeon Hollow were of several nationalities—Danish, Scotch, Irish, German, Holland Dutch, Welch, English. Most of them were second or later generation Americans who came from midwestern or eastern states. A few had lived for some years in California or Oregon before being attracted to the Palouse country. At that time in history "Palus" was a magic word. Palouse still is a magic word.

Centennial Saw Wheat Exports, Indian Trade

It was '76 and our country was 100 years old. Ulysses S. Grant was President. He appointed Mason Brayman Governor of the Territory of Idaho. Many regions of the west were settled and had been for years. The Palouse country was behind schedule and Pigeon Hollow was only getting started. However there was progress.

The first shipment of wheat was sent from the Palouse country to Portland. A farmer living near Colfax reported raising 1004 bushels of wheat sown on 30 acres of new ground. No history is recorded for the Hollow for that year of '76 but '77 was an eventful, lively year.

New boundaries were established for the Coeur d'Alene Indian Reservation and the northern, eastern borders of Pigeon Hollow became the southern, western borders of the Reservation. The Mission at DeSmet was begun in Hangman Valley.

Thereafter Indians regularly rode over the mountain and through Pigeon Hollow on their way to town. Farmington became their trading post. The Indians often rode home at night, whooping it up. No harm intended! The Coeur d'Alenes were friendly but sometimes their ways and their great curiosity disturbed the pioneers, especially the women.

The government granted a post office to Pine Creek settlement. Evergreen built a schoolhouse. Richard Truax's brother George planned and platted a new town which he named Farmington. Pigeon Hollow people sent their children to the school. They got their mail at the Post Office and enjoyed all the services to be had in a town close to their homes. The town grew and the services increased. The business of living eased a bit.

Two more men staked claims in Pigeon Hollow. Martin Hutch preempted a modest 40 acres in section 31. Mr. Hutch is remembered by at least one Farmington resident. John Irvin entered a timber claim for 160 acres in section 6 and became a neighbor to Mr. Warwick. No one in Farmington remembers Mr. Irvin but he and some members of his family are buried in Mountain View Cemetery.

How frustrated the Indians must have been when their trails were blocked by fences. Every farmer kept stock and every farmer needed a fence which he built of material available. In Pigeon Hollow it was wood. He built a rail or pole fence usually of pine. A rail fence would occupy a wide strip of land that could be put to better use. It invited a growth of weeds and brush. Snow would pile up before it. The fence was hard work to build and needed constant repair. The farmer needed a different type of fence to designate his boundary, enclose his crops and animals.

Barbed Wire Changed The Scene

In the east a search was on to develope a type of barbed wire fencing that would fill the homesteader's needs. The barbed wire must be of a material that would not sag in hot weather nor snap in freezing weather. It must be large enough so an animal could plainly see it and recognize it for a barrier. The barbs must be long enough to prick but not long enough to injure an animal charging against it. The barbs must not all slide to one place on the wire.

All during homesteading years inventors were experimenting with wire and they were busiest in the years when Pigeon Hollow was being settled. Finally several patterns of wire fencing were produced, any one of which met the Qualifications for a good enclosure. "The Winner" was best seller. "Baker Perfect" was second in popularity. "Haish S Barb" was also a good design and was offered for sale in Colfax at a very early date.

After the coming of the railroad spools of wire were available if a farmer had money to buy it. Wire fences replaced the rails but not in a day. Old rails were hauled to the house to be used for fuel. Neighbors would build and maintain a common fence. One old timer said, "You faced your neighbor across your boundary line and agreed. Each of us will build and keep up the half on his left."

The barbed wire fence proved to be practical and lasting. Nevertheless each farmer had to spend some time every year taking out weeds and brush, replacing old posts, stretching wire and otherwise fixin' his fences.

A three wire fence offered no great problem. One could crawl or roll under and hopefully perform the maneuver without becoming impaled on a barb. But a four or five wire fence-now that was a formidable barrier. To get on the other side one had to run to the nearest post and climb over, maybe or maybe not executing the climb and descent without getting hung up on any of the many barbs. If an angry bull were right behind you it was scary.

Fences were bothersome to drivers and riders. Every remote homestead needed its own road and Pigeon Hollow was laced with roads. It was the law that a man living near the main road could not deny access through his place to the man who lived beyond. Since each farm was fenced and even each field was fenced following a path through fields involved much jumping off and climbing on and opening and closing of gates.

One of these roads that were not roads was equipped with gates which had openers and closers. The driver had only tc reach out from his seat in the buggy, pull a rope and up went the gate; cluck to the horses, slap the reins, drive through, lean from the buggy, pull a rope and down went the gate. It was easy to visit here the owner in his cabin when he had such conveniences. What a tourist attraction it would be if the gates were still there!

Earl Poindexter says, "We had gates! One into Mr. Knight's field, a second into his hog pasture, a third into his barn lot, a fourth into his field, a fifth out of the Knight farm into the Delaney place and the sixth at the exit of the Delaneys." Going to town involved the opening and closing of six gates. For a child, climbing back on his horse after opening and closing a gate was no joke. We heard of one horse which was trained to put his head to the ground. The child sat on the horse's neck and the horse would lift his head sliding the rider into the proper place on his back. Not all horses were so trained. If the gate were made of wire there was no firm place to stand. A sagging with his weight barbed wire strand was a sticky support from which to clamber onto the back of a horse.

Just as the fences were slow in building they were slow to be taken out when no longer needed. One farmer said he started thirty years ago to take out fences and only now has completed the job. In a neglected corner of many an old barnyard can be found a coil or two of "bob wire."

Many of the fences and gates and homes are gone now. Homesteads were abandoned. The land was sold and consolidated into large farms. Some of the old roads that were not roads are now used to transport farm machinery to the "back eighty."

A Good Spring Makes a Difference

Charles Myers entered a homestead in 1882. He, at fifty was one of the older settlers. He had a young wife, Sarah and six school age children.

A quarter was staked by Henry Harlow in the same year. Henry and his wife Amanda sold the homestead and bought adjoining land in Evergreen. Why did the Harlows move across the fence onto land about the same as their homestead? One neighbor has said, "Perhaps they moved because the new place had a very fine spring." One cannot overestimate the value of a good spring. The Harlows lived in Farmington and in Evergreen the remainder of their lives. Their children were Lena and Park. Descendants still own the land.

The Palouse was fast changing its appearance. Indian trails were being cultivated out of sight and fields were taking the place of bunchgrass. Log cabins and fences and gardens showed where families lived. Even railroad reserves were often occupied.

The first long distance telephone line in Washington was strung from Colfax to Almota. Colfax saw its first train chug through town in '83. Grain began leaving on railroad cars to be shipped at Texas Ferry, now Riparia.[1]

1 From the Colfax Gazette.

Opal Ross

Gold was discovered in the Coeur d'Alene Mountains.

Farm wages were $1.00 a day but Chinese laborers on the railroad received 85 cents per day.

William Green Ritchey was born a pioneer. One night while the wagon train in which his mother was traveling camped on the bank of the Green River a baby was born to Mrs. Ritchey. She named the baby William Green. He was not the first nor the last baby to come into the world somewhere on the trail west. Frontier mothers were not allowed time to rest. The next morning the train moved on.

The young man, William Green Ritchey was living at Pine Creek when the Nez Perce Indians at Lapwai rose up to fight for their lands. William impulsively took off south to join a company of volunteers whose purpose it was to fight the Indians into submission. When he reached Lewiston he was surprised to hear that a white man had been killed by Nez Perce Indians at Pine Creek. He was even more startled to learn that the murdered man was his great uncle John Ritchey who had been living in a cabin in Pigeon Hollow. Now William had a personal reason for fighting. He joined the company with enthusiasm and vengeance. Later he wrote an account of his experiences in which he related the above story.

William Green and his brother John Ritchey were two of six who claimed Pigeon Hollow land in June and July 1885. The Hollow began to look like a settlement in its own right. Ritchey brothers and Irvine brothers settled the east central part of the valley. All four had been busy establishing themselves for several year and did not have to wait long for patents to their lands.

The Ritchey brothers were Scotch Irish. They were young men when they moved into the Hollow on neighboring places. William took over Uncle John's cabin and claim. He married Ina Bell Scott at Farmington. John married Rachel Irvine.

Landon and John Irvine were of Irish descent born in Virginia. John's farm was 80 acres. Landon got 160 in section 32. He planted his acres to orchard and timothy. He built a fruit drying and storage house, good barns and a milk house. An excellent spring supplied the water for this industry. Landon married Sydnah Ladd. There were six children. Landon and John are remembered by at least two Farmington people as Big Irvine and Little Irvine.

As the valley became occupied settlers moved up the mountain sides. Eventually fields would reach almost to the tops of the mountains. Any ground good enough to grow crops was used to grow crops even if it was steep. John Culp and William Click entered claims at the north and were the fifth and sixth to claim homesteads in the summer of '85.

Schools A Local Responsibility

Evergreen was building a new school and it was time for Pigeon Hollow to have a school of its own. Mr. and Mrs. John Cummings generously offered a tract of land on which to build. The legal document recording the deed reads:

"This indenture made the 4th day of October in the year of our Lord 1886. John and Martha Cummings of Nez Perce County, Territory of Idaho for the sum of $1.00 of the United States of America to them in hand paid by the trustees of School District no. 47 Nez Perce County release and convey unto the parties of the 2nd part and to their successors in office forever all the following described real estate Lot #5 section 6, T. 43, Range 5 W.B.M. commencing 35 rods from the S.W. corner of said lot thence north 13 rods thence east 13 rods thence south 13 rods thence west to the place of beginning, containing an acre of land, said above described land to revert to the said parties of the first part whenever said land shall cease to be used for public school purpose to have and to hold forever. John Cummings, Martha Cummings.
Signed-Samuel B. Ritchey, C. E. Pond"

The people built a plain little one room and cloakroom country type schoolhouse and shed for horses and rigs. There were no trees around the school but North Fork Pine Creek flowed nearby. Willows and brush grew on its banks.

Every morning teacher appointed two children to walk to the home of the nearest neighbor and bring back a pail of water. Teacher, pupils, visitors, all drank from the same dipper. At first the children sat on benches. Later they had desks, two by two. Generally small fry sat in front, big boys in the rear. For recitation at which time the children proved whether they had or had not learned their lessons each class in turn moved forward and sat on a bench facing teacher.

One teacher provided the education the children came so far to receive. For recitation at which time the children proved whether they had or had not learned their lessons each class in turn moved forward and sat on a bench facing teacher.

One teacher provided the education the children came so far to receive. Three or eight grades, one teacher! Miss Johnson taught the school for two years. She rode from her home in Evergreen and small brother rode with her. Earl Johnson recalls that Miss Johnson was no older than a high school girl.

Matt Johnson (not the same family) and the three oldest Poindexter children attended the Pigeon Hollow school. There is no attendance record and we do not positively know but surely Ritchey, Myers, Irvine, Smith and Harlow children and others must have been pupils there. After 1902 the school closed for awhile.

Latah County Born To Rich Resources

Nez Perce County to which Pigeon Hollow belonged was divided in 1888. The northern portion became Latah County and Moscow was named the headquarters for county government. All boundaries remained the same except the dividing line. Latah is a small county with some valuable assets. Its soil is deep and rich, some of the most productive in the United States. Its pine forests are famous. Its beautiful scenery is unsurpassed. The Farmington Register was proud of the town of Farmington and the country roundabout. The paper reported, "The wheat crop will be large this year ('88) as several hundred additional acres were sown." It was no doubt true. The paper went on to say, "The wheat yield is 40 bushels per acre. Wheat yields one year after another about the same." That may have been exaggerated a little.

Samuel Ritchey claimed the northeast quarter of section 32 and became a neighbor to his brothers John and William. In another two years Sam would marry Emma Cummings. Samuel Smith took the southeast quarter of the same section. He had a wife, Zerelda and a son Peter. Both Ritchey and Smith had been carving out farms before they made a declaration.

Neighboring territories of Montana and Washington became states in 1889. The following year the Territory of Idaho sandwiched between them also became a state. The three were numbered 41, 42, and 43 respectively.

1893 Year of Hardships

Growing season weather had been favorable and '93 crops were abundant. Farmers were happy in the thought of gathering a bumper yield. Harvest was in progress. Most of the wheat was still standing in the fields. Some wheat had been cut by headers and loosely stacked to await the coming of the threshing machine. A little wheat had been threshed and piles of sacked wheat stood in the fields. Then the weather struck a dreadful blow. Those who lived through that year never forgot what happened. They told the story to their children and the children remembered and repeated the story. It began to rain and continued to rain and kept on raining until the rain was replaced by snow. Standing wheat was a total loss. Piles of wheat were soaked and spoiled. Sack piles were bogged down in deep mud.

Chris Johnson was one of the luckier farmers. His wheat was threshed. He covered the sack pile with straw. When the straw was wet he removed it and spread on more straw. By using four horses on his wagon and hauling only a few sacks each trip he was able to pull through the mud and get the wheat to the barn. The damp sacks of wheat had to be shifted and turned all winter to prevent molding. Came spring and Mr. Johnson had something the world needed—wheat for spring seeding. He parcelled out the seed to his neighbors. A few could pay for it. Others had no money at all to pay for anything.

The new year brought no relief from hard times. Unaccountably the wheat market was depressed and wheat sold at 31 1/2 cents per bushel. Second grade wheat was 4 cents less. One Palouse farmer had to sell for 10 cents. The following excerpt came from a letter written in 1894 by F. M. Scott of Elberton,

"We had a market for our wheat; it was 10 cents to 20 cents per bushel. I sold 300 bushels of mine to buy something to live on. I got one sack sugar, 20 lbs. coffee, 40 lbs. dried fruit, 10 lbs. tobacco and $17.00 cash out of my crop so far. We make our own graham by running wheat through the chopper and make mush and gems out of it."[1]

In that year farm hands were paid $1.50 per day. Grain sacks were 10 cents each.

Chris Johnson was a blacksmith with an inventive mind. He bolted together two walking plows to create the first gang plow. Some of his property was virgin land covered with sun flowers. Chris found the new gangplow to be very effective in breaking up the tough sunflower roots.

Mr. Johnson and his wife came from Denmark. At the time of the '93 panic they were farming 160 acres in section 36. The children were five, Matt, Marie, Julia, Marion, Mae. Mrs. Johnson died. Chris sold his land rights and returned to Denmark taking his children. When he returned to Farmington a new wife came with him. They bought land adjoining the old place and lived in Pigeon Hollow. Dean Johnson was born to them.

Years of Prosperity

The effects of the panic were finally overcome. Wheat was up to 70 cents per bushel. Wages rose to no less than $2.00 up to $4.50 per day. The price of sacks came down to 7 1/2 cents each. A man could go to the woods for a large pine tree, pay $1.00 for the tree and take it home to make into fuel.

Mattie Headington was elected first Latah County School Superintendent. She covered her territory on horseback. With that type of transportation it is not likely she ever visited her far north district of Pigeon Hollow.

The government established a rurul free mail delivery service.

Beginning of the 20th century!

It was reported in the Gazette: "Wagon load after wagon load of harvesting machinery, grain sacks and binding twine line the country roads in every direction. Much new machinery has been bought."

Hungarian Partridges commonly called Huns were introduced into neighboring states and some quickly crossed the border into Idaho. Chinese Pheasants were turned loose in the State. The birds live in the grain fields feeding on gleanings and weed seeds. They are considered to be beneficial to agriculture. Both types of pheasants prospered and by 1916 hunting was allowed in Latah County.

Copies of the plat and tract books of T. 43 and T. 44, range 5 and 6 W. were filed in the Coeur d'Alene land office from the Lewiston office in Sept. 1901. In the future homesteaders could and did go to Coeur d'Alene to enter claims and prove up.

Mr. T.J. Kerl came to Pigeon Hollow with money to invest. Between 1901

[1]From the Colfax Gazette

and 1910 he bought large blocks of real estate, some of it farmland and some forest. He rented the farmland to several farmers. He increased his forest by planting fir trees. He built a summer home on a hillside near a good spring. Mr. Kerl's timber culture grew to maturity and the trees were harvested a few years ago by the present owner of the property, Mr. Schwartz.

Four men took possession of section 36. All declared in Coeur d'Alene on the same day, Aug. 22, 1903. N.P. Jensen and his brother-in-law, Abe Rasmussen claimed the east side. David Delaney and A.R. Knight got the west side for 172.9 acres each adjoining Washington. In the same year Chris Jensen staked 111.96 acres on the north ridge. The state line and the section line did not match. A narrow strip of land was left between them. These few acres became the home of the Chandler family.

N.P. Jensen who was popularly called Pete was born in Denmark. As a young man he came to Dayton, Washington and was joined there by Martina Rasmussen. They were married and two children, Roy and Mary were born in Dayton. The family moved to Pigeon Hollow and homesteaded. Clarence was born on the farm and grew up there. In the Hollow Clarence is the only descendant of a homesteader who still lives on and owns the original claim.

Many of the homesteaders listed here were not the first to farm the land. Often a man bought a squatters right and took over a place having a home and other improvements. He made his claim to the government land office and was entered in the books as a homesteader.

Pigeon Hollow properties were rented or sold over and over again. The list of owners becomes confusing. John Ritchey received the patent for his land in 1886, sold it a year later to Turpin D Anno who sold it in 1890 to John Culp who sold it six years later to John Torpey. One place was sold by E.G. Stonebraker to J.M. Merriman in 1892, sold again to M.W. Belshaw four years later, sold to R.C. Johnson in 1901, sold to T.G. Reid in 1908. Five owners in 16 years.

Homesteaders Largely Self-Sufficient

To produce food was the only way to earn a living in Pigeon Hollow. Every man was a farmer in his farm home surrounded by land more or less under cultivation, surrounded by fences, surrounded by chickens, backed by cows, pigs, horses and their habitations and watering troughs, backed by the chicken house, the backhouse, the smokehouse, the cellar, the wood shed, the garden, the orchard, the pump and well, all this not to mention the pigeons in the hayloft, the dog and the cat. A big spread supported one man and his family plus assorted dependent relatives. An early day good farm was an estate complete unto itself. If a farmer could sell a little surplus produce to pay taxes and a little more he could almost dare to feel secure.

A farmer obtained flour and chop for his animals by taking wheat to a mill to be ground. Some enterprising man might visit farmers soliciting wheat. If he could get the pledge of enough bushels he could afford to build a flour mill. Finally Farmington had its own mill. Pigeon Hollow and Evergreen farmers were saved many miles of hauling wheat to other towns to be ground into flour.

Winters of swirling, blowing, piling up snow, horses plunging through drifts desperately straining to pull whatever they were hitched to; children walking to school with their faces pressed against a biting wind, watering troughs frozen over, wet steaming clothes spread out to dry around the kitchen stove; shovelling paths to the barn, to the chicken house, to the outhouse, to the pump, even shovelling roadways; all these were facts of life which had to be faced and overcome by pioneer families.

Summer brought even larger problems—working all day long at the hardest kind of work on the hottest days, rubbing down the horses after the day's work was done, pumping water for stock, wearily milking the cows, and for the farmer's wife--skimming cream from endless rows of shallow pans of milk, washing on a board, ironing with a flat iron all the while keeping a lively fire going to heat the irons, taking care of sick children, planting a garden.

There were good times too--sleigh rides, coasting on homemade sleds (plenty of hills, both steep and gentle in Pigeon Hollow) long walks to the top of the mountain (lunch in pocket). Irene Belknap remembers the square dances in neighbor's homes. Earl Poindexter remembers the baseball games and horseback rides. Matt Johnson recalls sliding in the mud on the creek bank and driving to school in a buggy.

Threshing Season Brought Special Living

The T. S. Poindexter place lay in the northwest corner of the Hollow bordering Washington and the Coeur d'Alene Reservation. The piece contained 170.60 acres which Tom declared in 1906. A trail crossed their barnyard. Indians often went that way in going to or from Farmington.

Mr. Poindexter was born in Oregon. His wife, Sedalia, was a Palouse country girl. The children were Maxie, Arrah, Dean, Earl, Charlotte, Gilbert and Harold. Mr. Poindexter was owner and operator of a threshing machine.

Steam powered threshing machines were being used for harvesting in the early years of the century. This person knows very little about the working of a threshing machine except that a long belt with a twist in the middle like a tipped over figure eight joined together the separator and the engine. Power from the engine made the separator go. This person knows very little about the threshing machine but does know something about a cook house.

A cookhouse was a long narrow type structure like a room set on wheels and equipped with a tongue for the purpose of hitching on horses and going places with a half hour's notice. Inside the cookhouse were two long tables attached to the walls. Hollow box benches held the granite plates and cups and cooking pans and knives and forks. The lids of the benches were seats for the crew. At one end of the cookhouse was a giant iron cook stove that could put out heat like you wouldn't believe. The cook presided over this monster baking in the oven, frying and stewing on top.

The cookhouse was a shaky, shackey building but more sturdy than it looked. Each man arriving for a meal would grasp hold of the door frame and heist himself up noticeably jarring the whole building. Twenty-four men could enter and eat at one time. The cook and helper walked up and down the aisle between the rows of backs filling coffee cups, refilling platters and bowls.

The cookhouse was accompanied by a roustabout rig which was a rickety hack and two horses, past the best years of their lives. The roustabout had certain duties besides which he did everything no one else wanted to do. He rustled wood and water for the cookhouse, went to town or the ranch house for supplies, chopped the wood, took orders from almost everyone, took midmorning and mid afternoon lunches to the crew. Lunches were thick sandwiches stacked in a dishpan and covered with a towel.

The machine moved often and where it went there went the cookhouse. The roustabout helped move, piled wash pans, wash tubs, wash board, cook's tent etc. into the cookhouse and tamped it all down to stop the rattling as much as possible. When the whole loose package was ready to roll the boss ordered a teamster to hitch on and move it. The new site for the cookhouse had to be nicely chosen to be level; to be near enough to the machine so the men wasted no time coming to meals; to be far enough away from the machine so the chaff would not sift in. The machine moved from field to field, from farm to farm harvesting each man's crops in turn. The cookhouse and roustabout rig were as necessary to the harvest operation as the machine or horses.

Last Pigeon Hollow Land Claimed

After the season was over and the machine had pulled in it was traditional to serve chicken dinner to the crew and any others who wished to experience what it was like to eat in a cookhouse. The dinner was a grand finale to the harvest season.

In 1909 Franz Maureschat claimed a hard scrabble 84 1/2 acres. His land straddled a rocky north south ridge that seemed to divide the place in two. Franz lived on the east side and made his living with an orchard, a garden and pigs. He had no means of transportation. When the pigs were ready for market he drove them to town. Franz was a German. During the 1917-18 war when people were often hostile to Germans he stayed close to his ranch.

It was on the Maureschat place that the earliest surveyor found the "Indian trail which bears northeast, southwest." It was there the surveyor found "the sandy slope to the south and west." From anywhere on that slope the scenic view is positively superb. The east hillside has pine trees and remnants of the old Maureschat orchard. The hiker explorer strolling through might spook a deer or a coyote.

Fields and Pine Trees

Mr. Virgil McCrosky recognized the beauty of the old homestead and bought it to add to his park. A driver may turn off the Skyline Drive and follow the faint print of an old road down the ridge till he comes to farmland below.

One strip and two small corners of Pigeon Hollow land on the ridge in the timber remained to be taken. Judson A. Clark got the strip. Edward Jones took one corner of 19.62 acres. His patent was granted in 1917 ending the homesteading era in the Hollow. Virgil McCroskey got the last small parcel. He paid $11.90 for 11.90 acres which he needed to round out the outline of his park. Thus as late as 1943 Virgil's name was entered in land tract books at the end of the roster of homesteaders.

In the teens of the century cars were more and more coming into use. City people, farmers too or especially farmers took to the road. This development simplified life or caused life to be more complex depending on your point of view.

Henry Ford produced his Model T. It was of simple construction and could be repaired with hay wire. The first Model Ts were sold in 1908 at $850.00. Between then and 1924 the price decreased to $290.00. The car was so cheap almost every wage earner could have one. By 1918 half the cars on the road were Ford's Model Ts. The car was dubbed "The Tin Lizzie." Jokes about the Tin Lizzie were everywhere. Henry Ford considered the nickname and the jokes to be among his valuable assets. Oh, yes, the Ford could be driven up to 45 miles an hour if the driver and passengers could stand the rattling and shaking which was not all due to the car but could also be blamed on the roads.

School Time Again

About 1911 the Pigeon Hollow school board hired a teacher and opened the door of the old school. Not long afterward the school house burned. Children and teacher established themselves in a vacant house standing where the Albert Bruce home is today.

The district soon built a new school. It was a stout little building with a nice trim under the eaves. Pigeon Hollow people were proud of it. The well and pump were acquired a few years later.

All over Idaho there was increased interest in schools in 1914. Rural schools shared in the period of progress. Former pupils remember that year best and why not? They were in a new building with many companions and bright prospects for the future.

Roy Torpey remembers that Pigeon Hollow had 33 pupils in '14. Roy and Ruth Divine were seventh graders, Clarence Reid and Clarence Jensen were second graders. Mr. Schlaugh was the teacher. His pupils thought he was a very good teacher. There were so many classes that seventh graders, Roy and Ruth often had to hear the lessons of the two Clarences's in second grade.

Irene Belknap did not go to Pigeon Hollow school but the remainder of her

Opal Ross

Pigeon Hollow School, 1928-29
Pictures by courtesy of Thelma Simmons.
Left to right, Jack Belknap, Thelma Parsons, Helen Perkins.

family did and Irene remembers how it was. Harold, Marvin and Melvin Belknap were in school in '17 and '18. Other children in attendance were Clarence Jensen, Geneva, Thelma, and Herb Brooks, Mary and Maurice Torpey, Ronnie Bradley, Art Devine.

T.J. Torpey, clerk of the school board took a census count in Pigeon Hollow in 1920 and recorded the report in the courthouse. There were 36 children between the ages of six and twenty-one. Twenty-seven were definitely of school age--eight to sixteen years old. Twenty eight parents were listed. The next year's census counted thirty-nine children of school age.

Frances Torpey attended the little school as did Muriel Belknap. Muriel remembers her teacher, Miss Rupert.

The Parsons family lived on the road that goes over the mountain to the east. The Belknap family lived on the Mission Road. Thelma Parsons and Jack, the youngest Belknap walked part way home together. Jack was a little fellow and became very lonesome when he had to walk on alone. A few of the teacher's names are remembered by Thelma Parsons Simmons-- Miss MacDonald, Miss Helfert, Sophia Delagons.

Pigeon Hollow school had another turn in its long and varied career. In 1928 there were only three in attendance. The school closed. Then in 1934 the school opened again and had a seven year run. Jeanetta Schwartz finished her eighth grade and went to Potlatch to take a final examination. At that time the test was required of eighth grade graduates in Idaho.

Since about 1941 Pigeon Hollow children have been taken by bus to Farmington or Oakesdale schools. The schoolhouse was bought by a neighbor and moved to his yard where it still is. The pump and well are left in a field.

Nez Perce County Library has a books-by-mail program which is shared by Latah County residents. A borrower may ask the library for a book and it will be mailed to him with return postage and label included. The service is a fine supplement to the educational system of the counties.

The Legend of Pigeon Hollow:

Louie Sam was a Coeur d'Alene Indian whose personality and goings and comings in Pigeon Hollow and Farmington became a legend. Louie more than any other Indian from the reservation befriended the white people. Everyone had a story about Louie to tell and tell again.

His alloted acres were on the other side of the mountain north from the Hollow. There he lived surrounded by relatives, each on his own place. The old Mission road was near Louie's home and he often took to the road on his way east to the Mission or south over the hill to Farmington. He drove an old hack and two rangy cayuses. He wore a broad black hat and braids over his shoulders. His wife was often with him. Dogs trotted under the hack.

Prayer Chief of Coeur d'Alene Indians

"Prayer Chief" is the title of Louis Sam of the Coeur d'Alene Indian tribe, a title he upholds with all the dignity at his command. Among Indians at the De Smet mission, it is his honor to be first to enter the church, and to lead in prayers and singing. He is one of the most respected members of the tribe. (Photo by Edgar Dowd, S. J.)

Mrs. Clarke Torpey remembers that Louie Sam and his wife stopped occasionally as they went by on their way to town. Sometimes it was to buy fruit or to eat a meal, or to inquire about a person he hadn't seen for some time. He always was sad when we told him of the death of one he had known and usually remarked, "He was my friend." Mrs. Torpey recalls, "Louie Sam was an honest person and could be trusted."

Louie was greatly influenced by the teaching of the fathers at the Mission in DeSmet. It was his honor and privilege to be a Prayer Chief at religious ceremonies.

Many pictures were taken of Louie. The last pictures show that he gave up his braids and had his hair cut in business man's style.

The present chairman of the Tribal Council of Coeur d'Alene Indians, Bernard La Sarte is a direct descendant of Louie Sam.

Quite a few years have gone by since Louie drove his hack into town looking for friends to visit but even today in this part of the country dropping the name Louie Sam will spark a lively round of stories.

New Times, New Ways

Beginning in 1933 Congress enacted a series of laws to overcome the effects of the depression. The agencies were called alphabetical because they became known by their initials, usually three initials. The Agricultural Adjustment Administration (AAA) was intended to raise farm prices through the curtailment of production, the regulation of marketing and other devices. County committees made acreage allotments. Local committees were in direct contact with farmers. State committees were appointed by the Secretary of Agriculture.

It was difficult for farmers to curtail production when their great pride was to produce as much as possible. By planting wheat on their best land and by using fertilizer they contrived to raise larger and larger yields while still observing the program of the AAA.

The agricultural administration continued all the years unto the present time. The policies were adjusted to meet the needs of the farmers for each year. The program was expanded to include conservation practices to stop erosion. These included contour farming, windbreaks, cover crops, rough plowing for winter months and other methods. The agency had some name changes. At the present time it is called ASCS for Agricultural Stabilization and Conservation Service.

Clarence Jensen is a Pigeon Hollow farmer. He was elected to serve on a local committee of AAA at the very beginning of the program in 1934. Clarence continued to be elected to local and county committees. He was appointed to the state committee and traveled to Boise where he lived two winters in line of duty. Mr. Jensen was then sent to Omaha, Memphis, Chicago, Portland and Washington D.C. representing Idaho in the agricultural service. He has the distinction of having been with the agency longer than anyone else in the United States. Last winter Clarence was honored at a dinner and citation meeting in Moscow. He received two plaques which read:

"Clarence Jensen--Latah County ASCS office, Moscow, Idaho--For dedicated service to farmers and ASCS farm programs, for serving as either Community Committee, County Committee or State Committeeman each year since 1934 to Dec. 1974, forty years of service. As far as is known the longest service—as long as possible."

The second citation reads, "Plaque for Clarence Jensen. In grateful recognition of forty years service to ASCS, Agricultural Stabilization and Conservation Service."[1]

Forty years ago another alphabetical agency began operation. The (REA) Rural Electrification Administration made loans for the construction of lines to rural homes. Before that time only ten per cent of farm homes had electricity. Those who had it paid dearly to private companies for the service. Since that time 99 per cent of rural homes and businesses have been supplied with power through the work of REA.

Clearwater Power Company covers eleven counties in Idaho, Washington, and Oregon. One of these is Latah County. The far flung country lines are not easy to take care of especially in winter. But the service has always been good.

In 1938 construction crews moved into north Latah County and began to set up the system. Pigeon Hollow and Evergreen farmers had their first chance to observe the great American custom known as "the coffee break." People were scandalized to see the workers depart their jobs twice daily to drive all the way to Farmington for coffee. A farmer broke *his* fourteen hour day only once--at noon when he went home to dinner. The crews got their work done anyway and done well. The poles were set and the wires were strung. Pigeon Hollow homes were lighted in November.

The minimum charge for service was $3.30 per month. Some families doubted if they could afford to pay it. Almost all did wire their homes and people found electricity to be one of the best conveniences they had. And cheap. Nowadays when the current is switched off for a few hours to allow repair of the lines a housewife has a very helpless feeling. Most of her equipment is powered by electricity. She can hardly find any work to fill her day until the magic power is humming over the wires again.

Clearwater Power Company is a cooperative and once a year stockholders meet in Lewiston which is Company headquarters. The Company provides dinner in some luxurious place. The past two years dinner has been served in the Elk's Club overlooking the Snake River. After dinner comes the meeting at which reports are made and future plans of the Administration are revealed to stockholders. It is a day to have fun and meet acquaintances and other people.

Mr. Virgil McCroskey bought quarters and parcels of Pigeon Hollow land for his State park. This portion of parkland is a thin crown on the ridge and

[1]Information from ASCS office, Moscow, Idaho

only a small part of the vast park which extends south and east into Idaho for more than four thousand acres. McCroskey's Skyline Drive begins in Washington, curves high on the mountain over Pigeon Hollow, crosses the Old Mission Road at the summit of the pass and goes on for twenty-two miles till it reaches highway 95.

The park is a gift to the people for their recreation. In the last few years snomobiling is perhaps the biggest use of the park. In the park snow falls early, becomes deep and lasts a long time. Snomobilers are not allowed on roads but the park is theirs. Many of the enthusiasts unload their snomobiles at the Johnson ranch and leave their trucks in the yard while they race to the top of the hill night or day.

Mr. McCroskey set aside forty acres of parkland in section 29 for a game preserve. Native pheasants, deer and sometimes elk live there and all over the park. It is doubtful if hunters honor the boundaries of the game preserve.

Palouse Country Crop Specialties

In the Palouse country wheat is the thing. Hardly a day goes by without the thought crossing a Palouse farmer's mind, "What is the price of wheat today?" When a group of farmers meet sooner or later the conversation zeros in on the wheat market. A business man can join the group with zest and vigor. He thinks about it too because his prosperity revolves around the well being of the farmers. They all think about it and speculate but never can understand the ups and downs of the market, can never guess the best time to sell.

The lowest price we know about was in January 1894 when wheat sold for 10 cents to 20 cents and buying was not brisk even at that price. By 1896 the price was up to 70 cents per bushel and farmers were happy. Four years later wheat was down to 57 cents. "Between 1920 and '21 the price of wheat rose 25 cents a day, up and up while farmers held their grain for one more day and one more raise. Then when the price neared $2.50 the market started dropping by 50 cents a day and those who had held their crops were left with an almost worthless commodity.[1]

1925 wheat prices varied from $1.18 to $1.40. Then came the depression and the bottom dropped out. The McEwen Grain Company manager states that the lowest price they ever paid for wheat was 22 cents in 1932. Never mind whether that was a Portland or a Colfax price. If the farmer had to take the cost of shipping out of the 22 cents he had just as well keep the grain on the farm.

McEwen Grain Growers further state that the highest price for wheat was $6.00 in February 1974. May delivery for that year was $6.46. Some farmers had already sold and lost the benefit of the very high price. By February 26 the price was slipping and by March 5 it was all over. Buyers were not buying. By May 7 wheat was $3.50 and buyers were running. A farmer's life can be very exciting.

[1]Colfax Gazette

Opal Ross

Farmers had a good scare in the year '75. When harvest time rolled around it was raining. Combines and workers were ready and eager to go but it was raining. Some people recalled the stories they had heard about 1893 and wondered if it could happen again. It did not happen again. The normal harvest season passed. The rain stopped. The work was done later than usual. Everybody took a loss on grades but no one lost a crop.

Lentils are exclusively a Palouse country crop and for a long time these small seeds were planted only near Farmington. In 1916 a traveling German minister gave Mr. Jake Wagner a handful of lentils saying they probably would be valuable for food. Mr. Wagner planted his seed in an orchard between trees and among strawberry plants. Each year he planted a larger lot of lentils. He developed a mail order business and mailed little cloth bags of lentils to relatives and friends.[1]

In the fifty-nine years since that small beginning lentil production has increased to 100,416,000 acres in 1975. Pigeon Hollow farmers plant their share of these acres. Soil and climate in the Palouse country are just right for lentils. All are grown in a 25 to 30 mile wide belt along the Washington, Idaho border between Spokane and Genessee. An average yield is 1000 lbs. but farmers have produced up to 1500 or 1700 lbs. per acre. In 1973 the price of lentils reached an all time high of $30.00 a cwt. At the present time, second week in January, 1976 lentils are $10.90 a cwt.

In this part of the country vegetables are raised not to sell but to be used in the home or given away. Pigeon Hollow has some green thumb gardeners. Their vegetables and flowers could take first prizes in any county fair. Gardens and gardening provide a wonderful subject for conversations.

Community Relations Expand

Pigeon Hollow is part of the great Inland Empire. In your imagination or on a map place one end of a piece of string on Spokane and hold it down with your thumb. Extend the string to reach to the Columbia River. Make a sweeping clockwise circle. Take a swathe of British Columbia, a piece of Montana, all of the Idaho Panhandle, all of eastern Washington and a portion of northeastern Oregon, completing the circle where you started. The vast area within the circle is the Inland Empire which holds everything a man could ask of his environment, even space enough.

Community services for the Hollow are centered in many places. Farm products are marketed through facilities in Washington State. The school district is consolidated with Potlatch. North Latah County road district maintains the roads. Landowners are taxed a small millage for the upkeep of Mountain View Cemetery which is in Washington bordering the Idaho line. Julius Schwartz represents the neighborhood on the cemetery board.

Oakesdale-Farmington Fire District protects the community and

[1]Information from Idaho Pea and Lentil Commission and Miriam Schoepflin, Jake Wagner's daughter.

landowners are taxed for the protection. In case of fire a Farmington volunteer crew would respond.
If the fire were serious Oakesdale trucks would be called also. Clarence Jensen, August Wagner and Earl Palmer of Evergreen are the fire district board. A special telephone number is provided should a forest fire occur. It is a necessary service in the Hollow where some families live near the forest.

From 1908 to '58 Pigeon Hollow had a rural telephone system. All homes shared the line and a person could rubber if she wanted to. To "rubber" as any old timer knows is to hold the receiver to your ear and listen to the conversation of two other people. It was not all that bad. Often listening on the party line served a useful purpose. By rubbering a person could hear of sickness in another home or learn about the condition of the roads or hear of emergencies and could perhaps help. The telephone system is now owned by General Telephone Company of the Northwest with headquarters in Coeur d'Alene.

New Opportunities In Old Settings

Pigeon Hollow is no longer a bustling, well populated community. No children stand at the road in the morning to catch the school bus. The bus would make the run if only one child needed a ride. This year, 1976 for the first time there are no school age children in the neighborhood. Eight and part time nine families live in the Hollow.

Clarence Jensen is longest in residence having lived there all his life. He and his wife Mildred own and farm the old homestead and additional land. They have a daughter, Joan.

Julius and Josephine Schwartz came to live in Pigeon Hollow in 1931. They bought land and later built a new home. Their daughter, Jeanetta went to school in the little schoolhouse and in Farmington. Julius' family were Germans who migrated to Russia. Julius was born in Russia and came to North Dakota when he was a baby. Josephine's parents were also "Germans from Russia." Josephine was born in North Dakota. The Schwartz place has two good springs, one at the home and one on the hill in the timber.

In 1938 Lawrence and Nettie Johnson bought land in section 32. They found there several good springs, an orchard and four sets of buildings where other families had lived. Nettie's father was a German born in Germany. Lawrence is German and English. Their son Ronald grew up on the ranch and went to school in Farmington and Oakesdale.

August and Lillian Wagner, their son Arthur and his family live in the southeast corner of Pigeon Hollow. From 1943 to '75 the Wagners operated a dairy. Beginning with 10 or 12 Guernsey milk cows they increased and improved their herd. Much of the time they have milked 100 cows twice daily. The third home is that of the milker.

Lawrence and Violet Felker Bruce have lived in Pigeon Hollow for twenty-eight years. Their land is part of the Richard Truax homestead. Their home is next to the state line. The children, Lyle, Norma Lee, Janet went to school in Farmington and Oakesdale. Beverly attended only the Oakesdale school.

Albert Bruce Jr. and Ilene Pennel Bruce came to Pigeon Hollow to live after their marriage about twenty-six years ago. The children, Lynn, Douglas and Dan grew up on the farm and rode the bus to Oakesdale to school. Two years ago the Bruces built a fine new home.

Richard and Pamela Persons Pfaff and their two small children are late comers to the Hollow having lived there only about four years. If the Pfaffs continue to live where they are the school bus will be making the run again when the children are a little older.

Many, many families have lived in Pigeon Hollow both as renters and owners. It would be very difficult to trace all of them. I can name a few I know or know about.

John Cummings was one of the first pioneers to arrive in Pine Creek. He was born in Scotland, his wife Martha in Virginia. Their four older children were born in Kansas and Oregon. The last two children arrived in Idaho. One of these was Nellie. Mr. and Mrs. Cummings gave a tract of land to the district for a school.

Nellie Cummings married John Torpey. They bought a home on the Mission Road in section 31. The children were Roy and Clair. Until recently Roy farmed the homeplace and other land acquired by the family or by him. John was one of the four Torpey brothers who came from the east to begin a new life in Idaho or Farmington.

After their marriage in 1913 Clarke and Eula Hart Torpey went to live on the ranch in Pigeon Hollow. Clarke's parents, Mike and Minnie Torpey were residents of Farmington. Frances is the daughter of Clarke and Eula. Frances and her family live in Olympia and Eula lives nearby. In answer to a question about the farm Mrs. Torpey says, "The well is artesian and the flow was strong enough that the overflow was piped to the barn and kept the tank full."

Walter and Minnie Peddicord had a home on the north hill next to Franz Maureschat. The Peddicord daughter, Helen was a talented pianist and provided some rollicking music for square dances and parties.

R.J. Bell and Janie with their children Minnie and Vinnie owned a place near the state line and near town. It is presently the home of Lawrence Bruce and Violet.

In 1908 T.J. Reid bought a quarter in section 31. Reid and his wife had a large family--Clarence, Floyd, Charles, Leslie, Katherine, Alathine, Curtis, Adella.

November 1917 Clifford and Leona Belknap established a home on the Mission Road near the foot of the hill. Their children have been named in this account.

Fields and Pine Trees

Thomas was the first of the Torpey brothers to come from the east and settle in Idaho. He married an Evergreen girl, Winnie Price and they bought a place on the south border of Pigeon Hollow. They divided their loyalties between the two districts. Their children were Katie, Mark, Charles, George, Mary and Maurice. The four Torpey brothers mentioned above were Thomas, John, William and Mike. William and his family lived in Washington.

Fairview School in 1908
Picture by courtesy of Harold Kern.
Boy on left, second row, is Jake Howell. Third from left, second row is Henry Kern. Others are unidentified.

Fields and Pine Trees

Fairview—The Highroad Panorama

Fairview added to Evergreen, added to Pigeon Hollow comprise a political unit first called Pine Creek, later called Evergreen, and now called Farmington Precinct.

The main road follows the foothills of the mountains that guard Fairview on the east. Whenever, wherever the driver faces west, Steptoe looms large and broad and squarely before him. If he could go in a straight line up and down over the hills for about 20 miles he would run smack into the Butte. The driver could climb to the top and look around. He would see a million acres more or less of rounded hills all blocked out into fields. When these acres are covered with grain that is turning gold there is no more beautiful sight in the whole world. The grain literally is gold. It would take a computer to add the number of bushels sold and the number of dollars paid for it each year. Even the number of acres sown is a staggering figure.

There is no better place to view the panorama than the high road in Fairview. The view is indeed fair and even bettern' that. Back of the road the mountains are covered with pines, firs and tamarack too. People used to have homes on the slope. The driver-hiker could take a rough country road through the draw and into the trees. There he would find a lived-in cabin or two or three. The dwellers would have a few fruit trees, would raise a garden, would keep chickens or a cow, would cut wood, some to sell. It was not a soft living but it was an independent way to live. The standard of the economy did not always support a means of transportation. The dwellers might have to hitchhike to ride to town.

The mountains are a firm barrier separating the east from the west. On the east side roads lead to Potlatch or Moscow. The west side is Fairview. This story will try to account for the people and land of Fairview.

Streams from the mountain flow into Silver Creek which in turn flows into the Palouse River.

A spur from the foothills swings around and across the south boundary. The Woody Grade, named for J. M. Woody[1] is the road over the top. Woody was a pioneer who led a wagon train from Arkansas and arrived in the Palouse country in October, 1873. There were four families and six single men in seven wagons. Plenty of land was to be had. The homeseekers fanned out and settled in both Idaho and Washington Territories. Their lives were part of early history and some of their names are on landmarks today. As far as I have been able to learn the only member of the wagon train to settle in Fairview was J. M. Woody himself. He was a squatter in the northeast corner of the school section. His

[1]Mrs. Eva Nelson.

Fairview School Dist. #25
Latah County, Idaho T.43 R 566W
Homestead Map 1876-1893

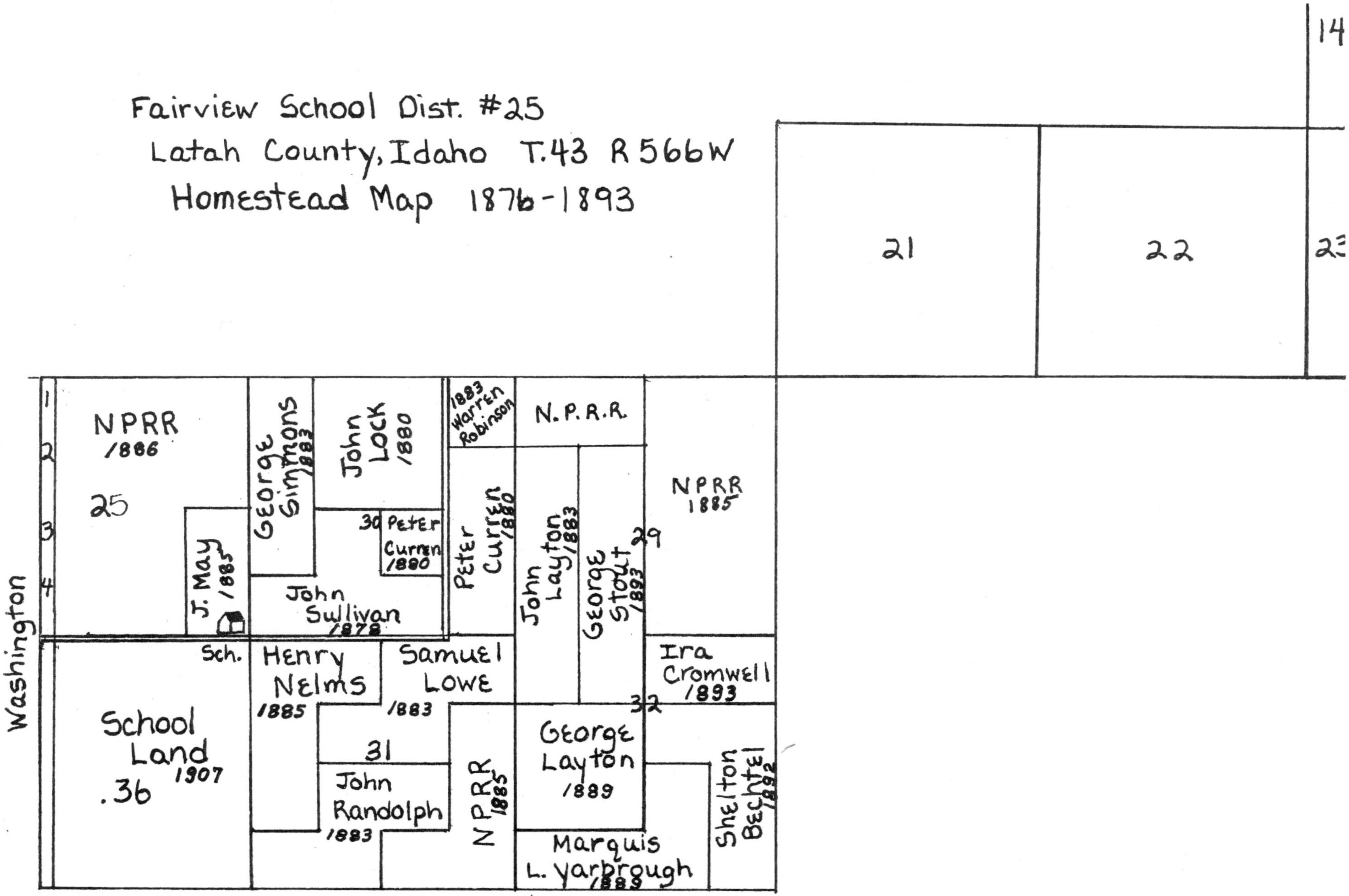

home was still there many years later.[1]

In winter snow drifts in from the south end, fills Woody Grade and stays there in shade until spring. Other times of the year the driver could travel the road, go over the summit, take any one of lots of roads and come to places with captivating names like Cora, Yellow Dog, Cedar Creek, Strong's Mill, Mountain Home and Freeze. Freeze was the name of two of the families in the wagon train. None of these places is marked with a signpost. The driver would have to know where he was or consult a map. As anyone could guess, young people who live in Yellow Dog take delight in telling new friends where they live.

Automobile owners used to drive out to Woody Grade to test the power of their cars. The grade was not all that steep but it may have seemed so to the driver of a 1918 model. Learning to drive and driving an early car was an adventure not to be forgotten. Grandfather Anton Futter ordered his first car through the mail and it was delivered to Tekoa. Taking his boys with him he drove a four horse team and wagon to Tekoa where he loaded on the car. At home Anton mounted the vehicle on blocks and set about learning to drive. No book of instructions came with the package. No one around could show him how. It was strictly a trial and error project. Starting the motor was an ordeal of cranking and cussing. The Frenchman pioneer finally got the car off blocks and on the road. The neighbors added bits and pieces to the story. For example· Once the car refused to stop and went through the store front into the store.[2] Anton wore out two or three of these not so durable models.

Settlers Contested By Railroad

The homestead map outlines the southern border of Pine Creek district. Two more sections on the east (not shown) are part of it. Section 21 and a quarter of Section 22 belong to Potlatch Forests. One quarter of section 14 and one half of section 22 is state land. Three quarters of section 14 and one quarter of section 22 belong to McCroskey Park.

Secton 32 does not belong to Pine Creek District but is added because two homesteads from 29 extend into section 32. It is also added because one of the homesteads belonged to George Stout and he is mentioned in this account and was well known in the neighborhood. Five sections remaining are Fairview School District.

The homesteading era began in 1878 when John Sullivan filed the first preemption for a quarter in section 30. John was a young man and a bachelor from Pennsylvania.

John Locke and Peter Curren claimed homesteads in the same section. Locke was 46 and had a young wife named Ida May, who was 21. John was a sociable chap and rode a white horse around the country stopping to visit with neighbors. He sported a beard. Eighty to eighty-five years after Locke lived on his homestead the present owners of the land cleared the ruins of his old home

1 Homer Futter.

2 Fundamentallly true, part fun.

and found a bottle labeled "whisker dye."[1] The two Johns, Locke and Sullivan formed a partnership and raised horses some of which they sold to the army.

The Northern Pacific Railroad Company fought with all the power it had to take the land promised to it by the government. The company did not always win. Sometimes it lost to the homesteader after causing him great anxiety. In Fairview the homesteaders were most often the winners.

Samuel Lowe and John Randolph went together to the land office to make their claims. On January 26, 1883, each applied for a quarter in section 31. Samuel Lowe was an Englishman who, with his wife, Annie, came to Pine Creek in 1876. They settled in Idaho Territory. Sam was respected by his neighbors. They said of him that he was well educated and could speak several languages. Sam and Annie had five children. Thomas, William, Arthur, Minnie, John. The boys farmed and operated a threshing machine. Tom was skillful in working with intricate mechanical devices. People took their watches and clocks and small machine parts to him to repair. The Lowe boys except one never married but they were fond of children and would always find time to give to a child. One or more of the Lowe boys lived continuously on the home place till Bill died in 1960. John married Ollie Slavin and they made a home nearby. Their daughter is Violet Lowe Strange.

Three homeseekers applied for land in 1883, George Simmons for 125 acres and Warren Robinson for 40 acres in section 30. John Layton took a quarter partly in section 29 and partly in section 32.

Indian Trail Still Shows

Henry Nelms laid claim to a place which had historical value. In pioneer days Indians were still following their established trails and stopping to camp and staying to play games and frightening homesteaders who never knew what to expect. Trails traversed the country and crossed and branched in a bewildering pattern. The surveyors found the trails but were too busy to follow and learn the destination of any one trail. And so the surveyor made a note in his field book, "Indian trail bears north east between sections 36 and 31" leaving us to wonder where the trail came from and where it went.

On Nelms place was a spring which may have been on an east west trail or a north south trail. Settlers observed that a trail went up the hill from Nelms' spring northeast. If the terrain permitted, Indians made a wide path not riding single file nor side by side but four, five or six riding roughly abreast.

Indians camped at Nelm's spring. On a flat nearby was a pond which made a fine center for a race track.[2] Exciting to think of the riders so long ago tearing round and round the pond, whooping their pleasure in the game, heels

1 Homer Futter.

2 Lavina Kern.

clamped tight, leaning into it, urging their ponies to go faster and win.

After breaking camp the riders went up the mountain. The path became so deep and firmly compacted that some years of cultivation were needed to erase it. All signs are blotted out now but the old trail has been found again. Where the trees and brush line join the field is a path perhaps twenty feet wide through a draw, not in the bottom of the draw where a tiny stream drains away the water, but beside it. The track is trampled and beaten deep into the ground. The plants growing there are different than the plants on either side.[1] These are characteristics of a trail which had long and steady usage but is no longer used. After two or three hundred yards surprisingly the trail is lost again.

James and Mary May staked 80 acres in section 25 and the Railroad Company claimed the remainder. The Mays left footprints for several reasons. They kept their land for a considerable time and sold space on the corner for a schoolhouse. Their son married a girl, Mary Jensen, whose family is well known in the area.

In 1889 homesteads were entered in section 32 by Marquis Yarbrough and George Layton. The Yarbrough land was first claimed by Randolph Heath and then cancelled. The place may have been bought later by the Heaths because they did live there at a later time.

George Stout took a quarter on the mountain. Ira Cromwell and Shelton Bechtel claimed the remainder of section 32. The school land was being used for pasture. After Will and Frank Futter bought the section in 1907 the spaces were all filled.

Pioneer Names Mostly Gone Now

In 1914 these families still owned their land--May, Locke, Simmons, Lowe, Curren, Stout, Futter. By 1930 all had sold out except Simmons, Lowe, and Futter. At the present time, 1976 only descendents of George Simmons still own the original claim.

George Kern and his wife were born in Germany. They came to Texas where George made a living hauling freight and building wagons. In 1890 the family journeyed to Idaho and bought the Sullivan land in Fairview. Henry was born on the place. There were other children in the family but Henry was the son who "carried on" the Kern family life in Fairview. He attended the original school and his memories as told to his wife and children are the only record we have of the old school. Henry was in the navy during World War I. He married a Garfield girl and they came to the home place to live after the death of the grandmother. There were four children, June, Joye, Harold and Donald. Harold is the son who "carried on." He and his wife, Lavina Hatley Kern,

[1] Harold Kern.

Opal Ross

Fairview School in 1932
Picture by courtesy of Harold Kern.
The teacher was Miss Underdoll
Left to right
Back Row: Mary Ellen Nelson, Harvey Weible, June Kern, Doris Nelson, Violet Lowe.
Front Row: Larry Culp, Ross Weible, Dick Culp, Milton Nelson, Roy Weible.

Fairview School
Picture by courtesy of Harold Kern.
The teacher is Mr. Charles Featherstone.
Pupil is June Kern.

bought the Lowe place where they presently live. Their son Johnnie is the fourth generation Kern to live in Fairview. Mrs. Ollie Kern lives in Garfield.

Anton Futter came from Alsace-Lorraine to Iowa. He and three of his brothers left Iowa to go round The Horn to Oregon. Many of the families who settled in Oregon were attracted to the Palouse country and Futters were among them. They moved on and arrived in Fairview. Anton rented and farmed the Nelms place and Indian land. Anton's sons Frank and Will bought the school land. Will built a home and lived on the south side of the section. His wife was Josie and their children were Cecil, Glen, Dora, and Verna.

The spot Frank chose for his home on the east side had been occupied by a squatter who cleared out. Futters found several drilled wells which were always a mystery. How the wells came to be there the family never knew. The well they used was a standing artesian with a good supply of water. Frank married Iva Geer and their children were Floyd, Jessie and Homer. After Frank's death Homer and his wife Glenna lived on the home place and farmed it. About 1945 they moved to Moscow where Homer became County Agent. He is now County Weed Supervisor.

Fairview Pupils Had It Good

Fairview had a very early school. As one man said it was "on the second hump over that way" which translated means north of the main road and north of the later and well remembered schoolhouse. Henry Kern went to school there as did the Lowe children and some Howell children. At some time before 1903 the school was destroyed by fire.

In 1906 Fairview district bought some land in section 25 for the purpose of establishing a school. Mary and James May sold for $50.00 one square acre more or less in the southeast corner of the section. The school that was built was directly across the road from the Kern home. Children who rode or drove tied up their horses in the Kern's barn. The children or teacher carried drinking water and washing hands water in pails from the Kern's home.[1]

The door of the schoolhouse was open for business for the next thirty-two years. The school saw a lot of history and many children were educated there. Of the teachers who are remembered, Charles Featherstone had the longest service. He was affectionately called "Chuck" and some of his pupils can't recall having any other teacher. Of those years Harold Nelson says, "Teachers at that time had a little right of way and could give a spanking if one were needed." Parents usually supported the teacher.

Fairview pupils had a loving care that was not afforded to most school children. They had a hot lunch. Mothers took turns coming to school to prepare hot cocoa or soup or something else hot. What a happy time the lunch hour must have been!

Margaret Bottjer, who was the teacher in 1908 gave each of her children a

[1]Ollie Kern.

Fairview School 1937-1938
—Picture by courtesy of Harold Kern
Fairview School District 25 in 1937.
The teacher was Miss Thelma Driscoll. The pupils were Harvey Weible, age 10; June Kern, age 10; Joy Kern, age 6.

souvenir program which listed the following attendance.[1]

Floyd Futter Elmer Smith Herbert Roberts Harvey Griner Eva Roberts
Carrie Smith Ernest Smith Beatrice Griner Ira Howell Henry Kern

Arthur May Meda Smith Oliver Griner Fred Griner

The School Board, S.J. Felts, Frank Futter, Henry Howell

Mary Jensen taught in 1911. Mary was a Pigeon Hollow girl and sister to Clarence Jensen. She married the neighbor boy, Henry May. Lou Ella Ross and Miss O'Hara were two other teachers. Miss Ross lived at Silver Creek. Lucy Horton Nocolai was the teacher in 1916 and 1917. She rode horseback or drove a buggy from her home between Cedar Creek and Silver Creek. Charlie Tipler was one of her pupils and rode with her. In 1934-35 17 children were in school and Miss Underdoll was teacher.

Boys and girls completing the eighth grade went to Potlatch to take state examinations required of them to earn a certificate of graduation. It was an exciting, even terrifying experience for a country boy to go to a strange school in a strange town and sit there amoung children who were strangers, biting his pencil, anxiously studying a long list of puzzling questions prepared by a learned scholar in the Superintendent's office, trying to remember what he had learned in the last eight years. Parents and teacher applied some pressure on him "to pass." [2] Girls, of course had the same problem.

The school was the center of a neighborhood with a large population. People met there for Christmas parties and the children entertained with songs and "Pieces" more or less well memorized.[3] June Kern says she has very dear memories of the pot-luck suppers, where Tom Stout and Alfred Weible would open a window, sit in it, and fiddle up a storm. Everyone would dance! All the neighbors came, and the small one room school was packed! Mrs. Alfred Nelson says wistfully, "People cared more for each other then. They shared and helped each other."

Miss Thelma Driscoll was the last to teach at Fairview. Her salary was $100.00 per month and the year was 1937-1938. Four were enrolled, Norman Ashley, Harvey Weible, June and Joye Kern. The boys moved away after Thanksgiving and then there were two. The school was closed and the children began to ride a bus to Garfield. Some years later the building was sold, torn down, moved away. The land reverted to the Nelson place. Following this history is a list of other children who went to Fairview not necessarily in the order of their attendance and not necessarily all of them--as many names as I could collect.

[1]Harold Kern.

[2]All eighth graders who talked to me.

[3]Milton Nelson.

Opal Ross

Old And New Share Today's Fairview

Grandfather Halvor Nelson was born in Norway. He and Martha acquired the first land owned by the family in Fairview when they bought the Nicolai place. Later Nelson's bought the May homestead. The son Alfred and his wife Eva Wilson Nelson moved to the ranch about 1919. Eva was a teacher from Kendrick, Idaho and of Scottish descent. There were five children- Harold, Walter, Milton, Mary Ellen, Doris. The grandmother and grandfather lived with them. On Sundays all went to Silver Creek Church. Halvor was 100 years, 19 days old when he passed away.

At the present time Harold and his wife Alberta Mesterman Nelson live on the home place. Alberta is Scottish and German. Their son Harold is the fourth generation Nelson to live in Fairview. Walter married Lenore Bruce of Farmington. Their home is in Washington bordering the state line. Milton married Margaret Evers of Garfield. Their home also is in Washington west of Fairview.

LeRoy and Janie Cocking bought the Futter place and came there to live in 1947. They have a farming operation and specialize in raising swine. When little piglets are arriving and finished pigs are departing for market any day of the week it is difficult to say how many pigs are on the ranch at any one time. Janie and her sons estimated there are 1000 all the time and the largest number at any one time was 1500.

LeRoy is English, Janie is a blend of nationalities and is proud that Cherokee Indian is part of her heritage. Tom, Jennie, JoDell, Scot and Steven are the Cocking children. Scot and Steve are the two still living at home. The Cocking home is partially surrounded by a colony of mobile homes. In one live Gus and Ila Elg who are friends to the Cocking family. Gus was born in Sweden. In another of the homes lives Rob Thonney. He has been there for twenty years and Janie cooks for him and takes care of his laundry. The third place is for the hired man. Frequent changes of occupants occur here. A herdsman lives with the family.

Diversity Added Interest

Any one of several stories might be suitable to be called "The Legend of Fairview." There is the morning glory story: A traveling salesman canvassed the district selling seeds. If a farmer bought $10.00 worth of seeds he would win a prize which was a package of flower seed. Two farmers qualified. They planted their seed which produced pretty flowers and were in fact morning glories. Every year the patches grew larger and there were more pretty flowers. There was no chopping them out. It wasn't funny.

Then there is the story of Alice Hill who lived on the mountain, her home hidden by trees. Alice would start out to walk to Farmington, at least seven miles away. Her three dogs would be with her, sniffing, exploring, loping on ahead, lagging behind, all four walkers straggling along almost together. A driver might stop and offer Alice a ride. O.K. Alice would be happy to ride but she wanted the dogs to ride too. It the driver didn't like to have three dogs in his car he didn't get Alice for a passenger. The four walkers stuck together and walked. Alice was a very bright woman and a competent teacher. She just liked dogs!

Alice once thought up a way to make some money and to make people happy at the same time. She organized the "Glad It Isn't Worse Medicine Company." The medicine was a packet of envelopes containing gentle sayings, cheerful thoughts, inspiring poems, and good wishes to send to convalescents, one envelope each day for a reasonable period of recovery of the patient. The idea was worthy of a longer run.

Then there was the story of Mr. George Stout who lived on his homestead far up on the hill. Mr. Stout would walk out to the main road and sit down to wait. When a driver came by going either way George would thumb a ride. He didn't care where he went. He just liked to go for a ride!

Then there was the son Tom Stout. He was a fiddler and he played very well. He was generous with his music and would willingly keep playing while neighbors danced the night away. Tom just liked to fiddle!

Little stories remembered: The hiker could climb part way up the mountain and find an old mine. The shaft is long and dark and silent.

It was fun for children to explore the tunnel with flashlight in hand, walk back into the mine, stumble over rough ground, splash in puddles of water, keep walking till they were overcome with terror and had to turn and run back into the daylight. Someone sustained by hope dug this long shaft with hand tools and maybe a little dynamite. As far as we know no ore of any value ever came from the mine.

A man named Bishop who lived near the Woody Grade invented a fanning mill which seemed to neighbors to be a good machine. Bishop made and sold some of the machines. He tried to get a patent and sell stock but was unsuccessful.

K.E. Holmes ran a steam threshing machine. Bill Tucker was water boy. Tom Lowe was engineer, Mabel Black was cook. Ole Dowell was separator man.

Alfred Nelson had a mania for tearing down old buildings. He and his sons took apart the vacant May house and used it to provide a home for Tom Stout after Tom's home burned.

Every farm had an orchard large or small. One was the Collins orchard. Spokane Beauty apples grew there. Mrs. Griner's trees were called "Mary Susan Orchard." [1]

Many Europeans countries were represented in Fairview — England, France, Germany, Norway, Sweden, Ireland, Scotland. Some of the first comers to the district were born in the old country.

[1]Harold Kern.

Opal Ross

The people of Fairview vote in a centrally located polling place in Evergreen. They receive their mail from the Farmington Post Office. Their roads are worked by the North Latah County Highway District. Their property is assessed a small millage for protection by the Farmington-Oakesdale Fire District. Their property is assessed another small millage for the maintenance of the Mountain View Cemetery. Their electricity is provided by Clearwater Power over R.E.A. lines. Their children are educated at the expense of Idaho but are sent by bus to Garfield, Washington. One home, Harold Nelson's is on the Garfield telephone exchange. The two other homes, Kern's and Cockings are on a Cora, Idaho exchange. Presumably the three families buy most of their supplies in Garfield for which they pay Washington sales tax. In Idaho they pay the property tax, a sales tax and income tax. How's that for a diversity of services and obligations?

School Census, Fairview District #25, 1920 All children 20 to 6 years of age
Charley Howell Alma, Sam, Violet, Gordon, Hugh, Arthur
James Kirkpatrick Clyde Middleton, Melvin Middleton
C.S. Smith Iva, Virgil
C. Ryser Rudolph, Fred, Rose, Anna
Will Futter Glen, Cecil

These were listed for 1924
W. Parr Wayne, Lyon
I. Pool May, Ben, Della, Fred, Maude, Lester
Sam Carter Gen
Chester Simmons Raymond
Frank Futter Jessie, Homer

Others
Alfred Nelson Harold, Walter, Milton, Mary Ellen, Doris
Henry Kern June, Joye
Alfred Weible Roy, Ross, Harvey
Harry Culp Dick, Larry
C.E. Jones Stanley, Fred, Dale
Frank Futter Homer
Hanson Dale, Gordon
John Lowe Violet
Bill Nolan Carol
Divine Mike, Virginia, Pat

Pigeon Hollow, Evergreen and Fairview were being settled about the time our country was 100 years old. It seems appropriate to name this history a bicentennial project and it is so named.

Opal Ross

Opal Ross

ACKNOWLEDGEMENTS

My sincere thanks to Charles Blickenderfer for allowing me to read his collection of family histories and other historical papers.

My sincere thanks to friends who shared with me their memories, their knowledge, their experiences.

Clarence Jensen
Eula Torpey
Earl Poindexter
Dean Johnson
Matt Johnson
Irene Belknap
Roy Torpey
Allie Torpey
Raymond Simmons
Thelma Simmons
Tracy, my husband
Tracy Ann, my granddaughter
Milton Nelson
Harold Nelson
Eva Nelson
Ollie Kern
Harold Kern
Lavina Kern
June Kern
Joye Kern
Homer Futter
Harry Duncan
David, my son
La Ree, my daughter-in-law
Kenneth Platt

BIBLIOGRAPHY

"The Nez Perce Indians and the Opening of the Northwest" by Alvin M. Josephy, Jr.
"The Conquest of the Spokanes, Coeur d'Alenes and Palouses" by B.F. Manring
"The Wire That Fenced the West" by Henry and Frances McCallum.
"Indian Thoroughfares" by Archer B. Hulbert
"The Coming of the New Deal" by Arthur M. Schlesinger, Jr.
Encyclopedia of American History by Richard B. Morris
Ruralite and Clearwater Power Co.
Brosman's "History of Idaho"
General Motors booklet
Pamphlets from the A.S.C.S. office
Brochures from Idaho and Washington Pea and Lentil Commission
General Telephone Co.
Father C.E. Byrne, S.J. of the Mission at DeSmet
History of North Idaho published 1903
St. Joe National Forest Service
Steve Agtee, Conservation Officer